Moritz Frahm

Occupant-Oriented Demand Response with Model Predictive Heating Control in a Multi-Zone Residential Building

Occupant-Oriented Demand Response with Model Predictive Heating Control in a Multi-Zone Residential Building

by
Moritz Frahm

KIT Scientific Publishing

Karlsruher Institut für Technologie
Institut für Automation und angewandte Informatik

Occupant-Oriented Demand Response with Model Predictive Heating Control in a Multi-Zone Residential Building

Zur Erlangung des akademischen Grades eines Doktors der Ingenieurwissenschaften von der KIT-Fakultät für Maschinenbau des Karlsruher Instituts für Technologie (KIT) genehmigte Dissertation

von Moritz Frahm, M.Sc.

Tag der mündlichen Prüfung: 17. Juni 2024
Hauptreferent: Prof. Dr. Veit Hagenmeyer
Korreferent: Prof. Dr.-Ing. Gerhard Schmitz

Impressum

Karlsruher Institut für Technologie (KIT)
KIT Scientific Publishing
Straße am Forum 2
D-76131 Karlsruhe

KIT Scientific Publishing is a registered trademark
of Karlsruhe Institute of Technology.
Reprint using the book cover is not allowed.

www.bibliothek.kit.edu/ksp.php | E-Mail: info@ksp.kit.edu | Shop: www.ksp.kit.edu

Print on Demand 2025 – Gedruckt auf FSC-zertifiziertem Papier

ISBN 978-3-7315-1385-8
DOI 10.5445/KSP/1000174319

Abstract

This thesis develops a novel framework for occupant-oriented demand response (DR) in a multi-zone residential building with distributed energy sources. The framework is based on model predictive control (MPC), optimizing the operation of a heat pump under consideration of photovoltaics (PV), batteries, DR, and time-variable occupants' thermal satisfaction (OTS) schedule. Two occupancy scenarios are assessed: baseline and adaptive, the latter being more complex and designed to reduce energy usage during unoccupied periods. Also, it is compared against conventional control strategies and examined in four different configurations, containing different combinations of PV and batteries. The results indicate that the new framework outperforms conventional control strategies, yielding more than 20 % in cost savings while also enhancing OTS. The implementation of an adaptive occupancy schedule reduces operational costs by more than 10 %. This can be further complemented by the integration of PV with more than 50 % additionally reduced operational costs. To identify various multi-zone model structures, black-box state-space identification algorithms are employed. The study includes three structures: central, decentral, and a new fusion approach. The three algorithms used are N4SID, CVA, and PARSIM-K. The PARSIM-K algorithm and fusion structure demonstrate the most consistent prediction accuracy with the fewest outliers and the lowest errors in the cross-validation. Validated over eight weeks of data from a real building, the MAE was 0.36 K, the RMSE 0.45 K in the short term prediction (24h) and 0.81 K MAE, 0.97 K RMSE in the long term prediction (168h). Overall, the new framework provides superior control performance compared to conventional control strategies. The framework is also highly versatile, allowing for easy integration of distributed energy sources and consideration of occupant needs. Furthermore, it offers efficient DR participation and employs an accurate, yet straightforward data-driven multi-zone modeling approach.

Kurzfassung

In dieser Arbeit wird ein neuartiges Framework für nutzerorientiertes Demand Response in einem Mehrzonen-Wohngebäude mit verteilten Energiequellen entwickelt. Das Framework basiert auf modellprädiktiver Regelung und optimiert den Betrieb einer Wärmepumpe unter Berücksichtigung von Photovoltaic (PV), Batterien, Demand Response und einem zeitvariablen Zeitplan für thermische Nutzerbedürfnisse. Es werden zwei Nutzer-Belegungspläne unterschiedlicher Komplexität evaluiert: ein Basisplan und ein adaptiver, wobei letzterer komplexer ist und darauf abzielt, den Energieverbrauch während der nicht belegten Zeiträume zu reduzieren. Außerdem wird es mit konventionellen Steuerungsstrategien verglichen und in vier verschiedenen Konfigurationen mit unterschiedlichen Kombinationen aus PV und Batterien untersucht. Die Ergebnisse zeigen, dass die neuartige Regelstrategie konventionelle Regelstrategien übertrifft, indem sie mehr als 20 % an Betriebskosteneinsparungen erzielt und dabei gleichzeitig die thermischen Nutzerbedürnisse besser erfüllt. Die Implementierung eines adaptiven Belegungsplans reduziert die Betriebskosten um mehr als 10 %. Dies kann durch die Integration von PV um weitere 50 % reduzierte Betriebskosten ergänzt werden. Zur Identifizierung verschiedener Mehrzonen-Modellstrukturen werden Black-Box-Zustandsraum-Identifikationsalgorithmen eingesetzt. Die Studie umfasst drei Strukturen: zentral, dezentral und einen neuen Fusionsansatz. Die drei verwendeten Algorithmen sind N4SID, CVA und PARSIM-K. Der PARSIM-K-Algorithmus und die Fusionsstruktur zeigen die beständigste Vorhersagegenauigkeit mit den wenigsten Ausreißern und den geringsten Fehlern bei der Kreuzvalidierung. Bei der Validierung über acht Wochen von Daten aus einem realen Gebäude betrug der MAE 0.36 K, der RMSE 0.45 K in der Kurzzeitvorhersage (24h) und 0.81 K MAE, 0.97 K RMSE in der Langzeitvorhersage (168h). Insgesamt bietet das neuartige Framework eine bessere Regelungsleistung als herkömmliche Regelungsstrategien. Das System ist außerdem äußerst vielseitig und ermöglicht die einfache Integration verteilter Energiequellen und die Berücksichtigung der Nutzerbedürfnisse. Darüber hinaus bietet es eine effiziente Demand Response-Beteiligung und verwendet einen präzisen und dennoch einfachen datengetriebenen Multizonen-Modellierungsansatz.

Acknowledgments

This thesis was written at the Institute for Automation and Applied Informatics (IAI) at the Karlsruhe Institute of Technology. With the completion of this thesis, I would like to thank everyone who accompanied me on my PhD journey and thus made a valuable contribution to this work.

Special thanks go to the supervisors of my thesis, Veit Hagenmeyer and Gerhard Schmitz, as well as my mentor Jörg Matthes. In particular, I would like to thank Veit Hagenmeyer for his great support in finding at a very early stage a research topic that suited my strengths and interests. In numerous, intensive technical discussions, we constantly refined my topic. I would like to thank Gerhard Schmitz for his great interest in my topic, the technical discussions, and his helpful feedback. I also thank Jörg Matthes for his excellent support, for stepping in as a mentor, and for providing a productive environment for my research. Thank you, Jörg for your strong leadership skills and constructive feedback, especially when we were writing publications.

I am grateful to have been at the IAI, which is characterized by numerous employees and a rich research infrastructure. In particular, I would like to thank the IAI staff of the Energy Lab infrastructure, among others, Simon Waczowicz, Luigi Spatafora, Jan Wachter, Philipp Zwickel, Tobias Moser, and Michael Walter. Thanks to them, I was able to carry out experimental tests. I would like to thank the FlexKälte project participants for the successful project work for the Federal Ministry for Economic Affairs and Climate Action, namely Heiko Maaß, Thomas Dengiz, Fabian Kern, David Wölfle, Wiebke Hofacker, Martin Wilferth, and Christoph Schlenzig. Special thanks go to Thomas Dengiz, who impressed me with his calm nature and particularly strong analytical skills. I am also happy about the fruitful exchange between the different research areas at our institute, especially with Ralf Mikut and his research area Automated Image and Data Analysis.

I would also like to thank the numerous doctoral students and undergraduates/graduates at the IAI who have taken the PhD journey with me or who have written their Bachelor's/Master's thesis with me, namely Felix Langner, Niels Jung, Pascal Strauch, Elena Klumpp, Vivien Geenen, Dominik Simeunovic, Anne Süß, Luisa Schorn, Stefan Meisenbacher, Moritz Weber, Chang Li, Miao Zhang, Dorina Werling, Frederik Zahn, Johannes Galenzowski, Rafael Poppenborg, Julian Hoffmann, Richard Jumar, Benedikt Heidrich, Rebecca Bauer, Jovana Kovacevic, Jakob Geiges,

Qi Liu, Max Kollmer, Andreas Döring, Katrin Reibelt, Markus Vogelbacher, and many more. I would especially like to thank Felix once again for being my most trusted companion on the PhD journey. Thank you Felix for our numerous, enriching discussions and collaborations. I would also like to take this opportunity to thank Stefan for his support in challenging situations. I experienced our coffee break sessions as incredibly refreshing and motivating, especially the digital sessions during Covid-19. Thanks also to Chang for being the world's best coffee group leader and for our Badminton training sessions.

Even if it is rather obvious, I am also glad to be able to write my dissertation in the age of the internet. Not only has it made it possible to start my dissertation despite Covid-19, but it has also enabled me to establish contact with international researchers. I would like to thank the IEA Annex79 participants for the enriching discussions. Most of all, I would like to thank Ján Drgoňa, who was always available to answer my technical questions about MPC, which has made it much easier for me to choose the right methodologies and tools. Also thanks to Weimin Wang for his fantastic proofreading of our collaborative publications.

Finally, I would like to thank my family. Special thanks go to my girlfriend, who has given me her full support at all times.

Preface

" *Aside from geothermal energy and lunar-induced tidal motion, the sun is essentially the only sustainable energy source available to species on Earth – whether directly via radiation, indirectly via photosynthesizing plants, or indirectly via the (ultimately sun-induced) wind.*"

Werner Nachtigal and Göran Pohl in *Bau-Bionik* (translated)

List of Publications

Journal articles

Bing Dong, Romana Markovic, Salvatore Carlucci, Yapan Liu, Andreas Wagner, Antonio Liguori, Christoph van Treeck, Dmitry Oleynikov, Elie Azar, Gianmarco Fajilla, Jan Drgona, Joyce Kim, Marika Vellei, Marilena De Simone, Masood Shamsaiee, Mateus Bavaresco, Matteo Favero, Mikkel Kjaergaard, Mohamed Osman, Moritz Frahm, Sanam Dabirian, Da Yan, and Xuyuan Kang. A guideline to document occupant behavior models for advanced building controls. *Building and Environment*, 219, 2022. doi: 10.1016/j.buildenv.2022.109195.

Moritz Frahm, Thomas Dengiz, Philipp Zwickel, Heiko Maaß, Jörg Matthes, and Veit Hagenmeyer. Occupant-oriented demand response with multi-zone thermal building control. *Applied Energy*, 347, 2023. doi: https://doi.org/10.1016/j.apenergy.2023.121454.

Felix Langner, Moritz Frahm, Weimin Wang, Jörg Matthes, and Veit Hagenmeyer. Hierarchical-stochastic model predictive control for a grid-interactive multi-zone residential building with distributed energy resources. *Journal of Building Engineering*, page 109401, 2024a. doi: https://doi.org/10.1016/j.jobe.2024.109401.

Felix Langner, Weimin Wang, Moritz Frahm, and Veit Hagenmeyer. Model predictive control of distributed energy resources in residential buildings considering forecast uncertainties. *Energy and Buildings*, 303, 2024b. doi: https://doi.org/10.1016/j.enbuild.2023.113753.

Conference contributions

Moritz Frahm, Elena Klumpp, Stefan Meisenbacher, Jörg Matthes, Ralf Mikut, and Veit Hagenmeyer. Development and Validation of Grey-Box Multi-Zone Thermal Building Models. In *BauSIM2022 - 9. Deutsch-Österreichische IBPSA-Konferenz: Tagungsband*. International Building Performance Simulation Association, 2022a. doi: 10.26868/29761662.2022.22.

Moritz Frahm, Felix Langner, Philipp Zwickel, Jörg Matthes, Ralf Mikut, and Veit Hagenmeyer. How to Derive and Implement a Minimalistic RC Model from Thermodynamics for the Control of Thermal Parameters for Assuring Thermal Comfort in Buildings. In *IEEE Open Source Modelling and Simulation of Energy Systems (OSMSES)*, 2022b. doi: 10.1109/OSMSES54027.2022.9769134.

Moritz Frahm, Stefan Meisenbacher, Elena Klumpp, Ralf Mikut, Jörg Matthes, and Veit Hagenmeyer. Multi-Zone Grey-Box Thermal Building Identification with Real Occupants. In *9th ACM International Conference on Systems for Energy-Efficient Buildings, Cities, and Transportation*, BuildSys'22. Association for Computing Machinery, 2022c. doi: 10.1145/3563357.3567403.

Moritz Frahm, Philipp Zwickel, Jan Wachter, Felix Langner, Pascal Strauch, Jörg Matthes, and Veit Hagenmeyer. Occupant-Oriented Economic Model Predictive Control for Demand Response in Buildings. In *Proceedings of the Thirteenth ACM International Conference on Future Energy Systems*, e-Energy '22, New York, NY, USA, 2022d. Association for Computing Machinery. doi: 10.1145/3538637.3538864.

Philipp Zwickel, Moritz Frahm, Johannes Galenzowski, Karl-Heinz Häfele, Heiko Maaß, Simon Waczowicz, and Veit Hagenmeyer. Demand response in smart districts: Model predictive control of building cooling. In *2022 IEEE PES Innovative Smart Grid Technologies Europe (ISGT-Europe)*, 2022. doi: https://doi.org/10.1109/ISGT-Europe54678.2022.9960608.

Contents

1 Introduction

Building operations account for 30 % of global final energy consumption and produce 26 % of total energy sector emissions (Internation Energy Agency 2023a). Approximately half of this usage can be attributed to space conditioning using heating, ventilation, and air conditioning/cooling (HVAC) systems (Internation Energy Agency 2023b). Optimizing HVAC operation presents significant potential opportunity to improve energy efficiency and increase occupants' thermal satisfaction (OTS) (Minoli et al. 2017).

Renewable energy sources can fulfill the electricity demand of buildings and lead to decreased emissions (Vieira et al. 2017). As these sources supply energy intermittently, HVAC operations in buildings can adjust accordingly. HVAC systems and other flexible electricity consumers can participate in demand response (DR) programs to handle the increasing share of renewable energy sources (Dengiz et al. 2019). DR refers to the changes in customers' electricity consumption patterns in response to flexible electricity tariffs (Albadi and El-Saadany 2007). These tariffs adapt to current market conditions, such as the availability of low-cost renewable energy and high prices due to energy shortages.

Improving HVAC operation is therefore undoubtedly crucial. However, it is essential to note that energy demand is directly linked to occupants' requirements for comfortable personal conditions (Hong et al. 2017, Janda 2011). Considering occupant behavior (OB) in the HVAC control-loop can reduce energy consumption (Pang et al. 2020) and improve OTS (Xie et al. 2020). For example, the average occupancy rates are usually below 60 % (Mahdavi et al. 2008), resulting in avoidable energy use during unoccupied periods. Masoso and Grobler (2010) observed that 44 % of energy consumption occurs during occupied hours, while 56 % is consumed during unoccupied periods.

A possible optimization strategy for HVAC operation involves implementing advanced control strategies, such as model predictive control (MPC) (Drgoňa et al. 2020). MPC solves an optimization problem by considering system dynamics, forecasts, and constraints, to determine the optimal input trajectory over a predictive horizon (Hazyuk et al. 2012a,b). It is well-suited for dynamic systems with rather slow response rates, such as those found in the thermodynamics and heat transfer of buildings (Hu and Karava 2014). In addition, MPC can integrate building models (Prívara et al. 2013), weather forecasts (Oldewurtel et al. 2012), OB (Oldewurtel et al. 2013), and provide DR (Oldewurtel et al. 2010). However, model development poses a significant challenge in the MPC design process (Cígler et al. 2013). Drgoňa et al. (2020) highlight the need for balancing model accuracy and simplicity as accuracy plays a crucial role in MPC quality, while complex models lead to increased computational demand. This is especially true for room-specific control, as the modeling process can be quite demanding (Atam and Helsen 2016).

1.1 Related Work

MPC for buildings and related subcomponents such as control methods, OB, and thermal building models have already been investigated by a number of studies (Drgoňa et al. 2020, Dong et al. 2022, Park and Nagy 2018, Naylor et al. 2018, Kaspar et al. 2022). Drgoňa et al. (2020) provide a comprehensive introduction to MPC in building control, covering different models, algorithms, and practical tools. Dong et al. (2022) evaluated OB models in literature and found a limited number of publications discussing the integration of OB into building control. Park and Nagy (2018) investigated the connection between OTS and building control research and determined that the majority of studies prioritize on energy savings over OTS integration. Naylor et al. (2018) reviewed occupant-oriented control strategies and found out that more complex control systems, such as predictive ones, yielded the highest energy savings. Kaspar et al. (2022) evaluated DR control schemes in the literature and reported that many studies reduce both energy costs and peak load while considering OTS. Nevertheless, they indicated a scarcity of practical applications. Overall, literature reviews focus on one or more of the following five subtopics in building control: control methods, OB, thermal comfort, DR, or thermal building models. However, a comprehensive analysis is still lacking, especially in connection with practical implementations and room-specific control operations.

Multi-zone thermal building modeling

Room-specific control operations can be addressed with multi-zone models. Multi-zone thermal building models have been reviewed by Shin and Haberl (2019) and Atam and Helsen (2016). Shin and Haberl (2019) conducted a review of thermal zoning definitions, revealing consistency among sources in defining thermal zones as spaces that are controlled by their own setpoint and schedule. Atam and Helsen (2016) provide an overview of control-oriented multi-zone modeling, describing various model structures, such as white-box, grey-box and black-box models, as well as central or decentral structures. According to Drgoňa et al. (2020), white-box models are based on physical laws, while black-box models rely solely on input-output data. Grey-box models combine both approaches by using data to identify predefined model parameters. Overall, grey-box and black-box approaches seem relevant, as they involve a comparison with real measurement data. Therefore, the remainder of this review will focus on comparing grey-box versus black-box methods as well as centralized versus decentralized structures.

Most multi-zone thermal building model (MZTBM) research employs grey-box models with resistor–capacitor (RC) circuits or black-box models that use subspace identification techniques like N4SID (Prívara et al. 2013, Vallianos et al. 2022, Drgoňa et al. 2021). Prívara et al. (2013) compared multi-zone grey-box RC models with N4SID black-box models under two different scenarios, one with simulation data and the other with real measurements. Their conclusion was that real, large systems are too complex for RC models, making N4SID the only feasible choice. Vallianos et al. (2022) identified multi-zone RC grey-box models from empirical data, achieving accurate predictions for up to 24 h in advance. Through a ten-day identification period using real measurements, Drgoňa et al. (2021) evaluated various black-box state-space multi-zone models, including MOESP, CVA, N4SID, and Neural State-Space models. They found the best outcomes in terms of low prediction errors with Neural State-Space and MOESP for ten days of open-loop testing. Overall, few comparisons between grey-box and black-box multi-zone models currently exist. None of these authors indicated whether they used central or decentral multi-zone modeling approaches.

Further analysis of grey-box and black-box models can be found in literature assessing single-zone models (Yu et al. 2019, Lin et al. 2012, Ferkl and Jan Široký 2010). Yu et al. (2019) conducted a study comparing a single-zone RC grey-box model with an N4SID black-box model using real data. Their analysis revealed that the N4SID approach slightly outperformed the grey-box model. Lin et al. (2012) compared single-zone grey-box RC models of different orders with real data. The model order describes the number of states used, e.g. the number of temperature nodes. They conclude that low-order models, even first order, are sufficiently accurate. They also stated that grey-box parameter estimation from closed-loop data leads to inaccurate parameter estimates. Ferkl and Jan Široký (2010) evaluated two single-zone black-box models using real data: N4SID and ARMAX. Both methods yielded results with standard deviations of less than 0.3 K, indicating their effectiveness. They noted that N4SID is more accurate for noisy systems and more convenient as ARMAX requires more parameters and is also more sensitive to parameter changes. In summary, there are only a few studies examining the use of black-box state-space models like N4SID in the building sector. The results indicate that its performance is comparable to other modeling methods such as grey-box RC and ARMAX. However, the handling of closed-loop identification data is still unresolved.

In existing literature, a comparison between central and decentral approaches exists only for multi-zone RC grey-box models (Arroyo et al. 2020, Joe et al. 2020, Frahm et al. 2022c, Freund,

S. and Schmitz, G. 2020). Arroyo et al. (2020) identified multi-zone grey-box RC models from simulation data, where they compared decentral and central structures as well as different data lengths. They noted that one week of data was sufficient to identify multi-zone building models. In their simulation-data-based study, the centralized model slightly outperformed the decentralized model. In contrast, Joe et al. (2020) identified central and decentral multi-zone grey-box RC models from real data from an unoccupied building. In this real data-based study, the decentral approach outperformed the centralized approach in terms of model accuracy. Frahm et al. (2022c) evaluated multi-zone grey-box RC models with real data in an occupied building and compared central and decentral structures as well as different algorithms and data from occupied and unoccupied periods. The information quality, specifically the occupied versus unoccupied distinction, had the most notable impact on model accuracy. The structural and algorithmic differences were comparatively minor. Freund, S. and Schmitz, G. (2020) identified multi-zone grey-box RC models of different orders, where low-order models yielding the most accurate and generalized outcomes. In summary, the results so far indicate that a low model order and high information quality without disturbances are most likely to improve the quality of grey-box RC models. When comparing central and decentral grey-box models, no clear preference in terms of prediction accuracy was found. A comparison between central and decentral black-box models has not been studied, yet.

Multi-Zone Control Simulations

While many MZTBM are identified from real data, most studies on control algorithms are performed via simulations. This analysis focuses on the existing literature on simulating occupant-oriented DR with multi-zone control. Two types of occupant-consideration are investigated: baseline and adaptive occupancy consideration. The baseline consideration assumes a constant schedule, e.g. during working hours from 8 AM to 5 PM, while the adaptive occupancy consideration accounts for more varied and complex schedules, taking into consideration the actual occupancy.

The adaptive occupancy schedules have been implemented by Huang et al. (2015), Pedersen and Petersen (2018), and Korkas et al. (2016). The latter two studies, Pedersen and Petersen (2018), Korkas et al. (2016), also simulate participation in a DR program. Huang et al. (2015) obtain occupancy data from a Johnson Control BMS for a real-world dataset and train a multi-zone artificial neural network-based thermal dynamics model. Pedersen and Petersen (2018) use occupancy estimates derived from real CO_2 measurements and developed a decentral multi-zone grey-box RC model. That model was identified from a ten-zones EnergyPlus model. Based on the occupancy data, they apply a rule-based control (RBC)-based optimal start–stop control strategy. Korkas et al. (2016) model three buildings in EnergyPlus with ten zones each and shift HVAC loads based on occupancy schedules. In their study, the same occupancy schedule is applied to each zone within the buildings. However, each of their buildings is operated by a different schedule. Overall, these studies demonstrated successful integration of real occupancy data.

Further investigation of multi-zone control can be found in research that employs simpler occupancy schedules (Blum et al. 2016, Mork et al. 2022, Hu and Karava 2014, Hu and You 2023, Biyik and Kahraman 2019). Blum et al. (2016) implemented a multi-zone grey-box room transfer function, identified from an EnergyPlus model with 18 zones, to assess the DR potential. They also studied the impact of thermal storage on the DR potential and discovered that structures with higher storage yield lower costs. Mork et al. (2022) evaluated three different multi-zone control strategies, namely central, decentral, and distributed, using a central Modelica model. Hu and Karava (2014) implemented a multi-zone grey-box RC model with time-variance. They applied it for a mixed-mode cooling control strategy, where the windows can also be actuated. Hu and You (2023) compared various robust MPC formulations in a multi-zone BRCM model to account for uncertainties in weather forecasts. The control-model is identified as a multi-zone grey-box RC model from BRCM. Biyik and Kahraman (2019) incorporated photovoltaics (PV) and a battery into the multi-zone control strategy. The multi-zone grey-box RC model was identified from real data. Instead of using an occupancy schedule, the authors implemented a constant predicted percentage of dissatisfied (PPD) model. In summary, these studies reveal the feasibility of applying multi-zone control to building control for a wide range of applications including investigating parameter studies, mixed-mode control, forecasting uncertainties, and PV self-consumption. It remains open to bring these aspects together.

Multi-Zone Control Field implementations

Compared to studies based on simulations, there is a scarcity of field-evaluated multi-zone control studies. Freund and Schmitz (2021) and Maddalena et al. (2022) implemented multi-zone control in the field, but without incorporation of DR. Freund and Schmitz (2021) identified a decentral multi-zone grey-box RC model from real data and implemented a multi-zone MPC strategy in the field while considering a simple occupancy schedule. In terms of energy use, the heating demand was reduced by 30 % compared to conventional rule-based control (RBC). In terms of comfort, the solar heat gains and heating system restrictions resulted in overly warm conditions. Maddalena et al. (2022) identified a black-box multi-zone Gaussian process dynamical models from real data and and utilized them in a real-world implementation of MPC. However, their proposed implementation in MPC led to non-convexity and lengthy solve times. Other studies on implementing MPC in fields are available for single-zone optimization (Finck et al. 2019, De Coninck and Helsen 2016, Široký et al. 2011). Finck et al. (2019) evaluated DR in a field study using a black-model single-zone artificial neural network model that was identified from real data. De Coninck and Helsen (2016) and Široký et al. (2011) both utilized MPC with grey-box models in a single zone, but did not include DR participation. In summary, the existing studies indicate that multi-zone control is appropriate for implementation in the field, yet. Further research is needed to unlock the full potential of MPC in buildings.

Summary

Studies about modeling usually involve a comparison between different model types, such as grey-box versus black-box or high versus low model orders. Typically, black-box state-space models demonstrate superior performance in accuracy compared to grey-box models. Nevertheless, grey-box models can retain physical interpretability in contrast to black-box models. Grey-box models have been the subject of comparative studies on the effectiveness of central and decentral approaches. Such comparison between central and decentral approaches is still lacking for black-box state-space models. Additionally, concerns remain about the use of closed-loop identification data, which is often applied in real-world scenarios.

Research on multi-zone controls investigates different aspects such as DR, forecast uncertainties, parameter studies, energy savings, occupant behavior integration, thermal comfort, or field tests. Most multi-zone control strategies are occupant-oriented, applying an adaptive or baseline schedule. However, to objectively evaluate the performance of room-specific occupancy, it is necessary to compare baseline and adaptive occupancy schedules, which is still lacking. Furthermore, only a few studies have assessed the integration of distributed energy sources such as PV and battery systems in multi-zone control. If thermal storage is already being utilized through DR, there is still a need to compare the additional advantages that batteries can bring.

1.2 Contributions of the present Dissertation

This thesis addresses the research question on *How to provide demand response in a multi-zone residential building with distributed energy sources while considering room-individual occupancy?*

Therefore, the present dissertation introduces a novel framework for occupant-oriented demand response with multi-zone thermal building control. This framework aims to fill the research gaps in modeling and control strategies identified in the previous section. The framework is founded on model predictive control (MPC) and accounts for photovoltaics (PV), batteries, DR, and a time-variable occupants' thermal satisfaction (OTS) schedule. The framework evaluation involves simulation studies that compare a baseline and adaptive occupancy schedule and assess uncertainties in model and weather forecasts. Various model structures are examined for the black-box state-space models, comprising central, decentralized, and a new fusion approach. Multiple identification algorithms for black-box subspace identification were investigated, including N4SID and PARSIM-K, the latter of which is specifically suited for closed-loop identification.

1.3 Dissertation Structure

Chapter 2 outlines the theoretical fundamentals for developing a new framework, including the laws of thermodynamics, thermal building models, OTS, and control theory. In Chapter 3, the methodology for the framework is presented, organized in submodules such as the model, consideration for OTS, control inputs/outputs, predictions, constraints, and optimization formulations. Chapter 4 evaluates the performance of the framework in terms of operational costs and OTS. This assessment compares the new framework with conventional control strategies while also evaluating different framework configurations. Chapter 5 provides experimental outcomes for the model identification and validation conducted on a real building. Lastly, Chapter 6 concludes the thesis.

2 Fundamentals

This chapter introduces the fundamentals required for the development and evaluation of the novel framework. It gives an overview of the necessary laws of Thermodynamics and heat transfer, thermal building models, occupants in buildings, and building control systems.

2.1 Laws of Thermodynamics and Heat Transfer

The laws of Thermodynamics and heat transfer are the fundamentals of the thermal building models, and thermal satisfaction models. The first and second laws of Thermodynamics are presented in the following, based on (Elsner and Dittmann 1993, Baehr and Kabelac 2016).

2.1.1 First Law of Thermodynamics

The first law of Thermodynamics defines the energy balance. It states that the change in energy over time within a system is equal to the energy transported across the boundaries of the system. Eq. 2.1 presents the first law of Thermodynamics for an instationary, open system (Elsner and Dittmann 1993, Baehr and Kabelac 2016).

$$\underbrace{\sum_i \dot{Q}_i}_{\text{Heat flows}} + \underbrace{\sum_j \dot{W}_j}_{\text{Work flows}} + \underbrace{\sum_k \dot{m}_k \cdot \left(h + \frac{c^2}{2} + g \cdot z\right)_k}_{\text{Energy transport coupled to mass flows}} = \underbrace{\frac{\mathrm{d}}{\mathrm{d}t}\left(U + E_{\text{kin}} + E_{\text{pot}}\right)}_{\text{Change in energy over time}} \tag{2.1}$$

- **Heat flow** $\dot{Q}$: Heat flows in form of conduction, convection, and radiation.
- **Work flow** $\dot{W}$: Work flows are electric energy, change of volume, and friction work.
- **Energy transport coupled to mass flows** $\dot{m} \cdot \left(h + \frac{c^2}{2} + g \cdot z\right)$: This form of energy is supplied or removed via mass flows.

- **Internal energy** U: The internal energy consists of internal thermal, internal chemical, and internal nuclear energy. As the characteristic state variable, it represents the sum of the energy of individual particles.
- **Kinetic energy** E_{kin}: The kinetic energy results from the movement of the center of gravity.
- **Potential energy** E_{pot}: The potential energy depends on the position of the center of gravity.

The changes in a system's kinetic and potential energy can often be neglected in thermal systems, such as buildings. For the neglection of the kinetic and potential energy, the first law of Thermodynamics can be simplified to Eq. (2.2).

$$\underbrace{\sum_i \dot{Q}_i}_{\text{Heat flows}} + \underbrace{\sum_j \dot{W}_j}_{\text{Work flows}} + \underbrace{\sum_k \dot{m}_k \cdot h_k}_{\substack{\text{Energy transport} \\ \text{coupled to mass flows}}} = \underbrace{\frac{\mathrm{d}U}{\mathrm{d}t}}_{\substack{\text{Change in internal} \\ \text{energy over time}}} \tag{2.2}$$

2.1.2 Second Law of Thermodynamics

The second law of Thermodynamics introduces the state variable entropy that quantifies the convertible restrictions of different forms of energy. Eq. (2.3) presents the second law of Thermodynamics for an instationary, open system (Elsner and Dittmann 1993, Baehr and Kabelac 2016).

$$\underbrace{\dot{S}_{\text{prod.}}}_{\substack{\text{Produced} \\ \text{entropy flows}}} + \underbrace{\sum_i \left(\frac{\dot{Q}}{T_{\text{A}}}\right)_i}_{\substack{\text{Entropy transport} \\ \text{coupled to heat flows}}} + \underbrace{\sum_j \dot{m}_j \cdot s_j}_{\substack{\text{Entropy transport} \\ \text{coupled to mass flows}}} = \underbrace{\frac{\mathrm{d}S}{\mathrm{d}t}}_{\substack{\text{Change in entropy} \\ \text{over time}}} \tag{2.3}$$

- **Entropy** S: Entropy is a state variable that changes with irreversible processes inside the system, heat transfer over system boundaries, and mass flows over system boundaries.
- **Produced entropy flows** $\dot{S}_{\text{prod.}}$: Entropy can only be produced, not destroyed: $\dot{S}_{\text{prod.}} \overset{!}{\geq} 0$. In a reversible process, no entropy is produced: $\dot{S}_{\text{prod.}} = 0$.

- **Entropy transport coupled to heat flows** $\frac{\dot{Q}}{T_{\mathrm{A}}}$: This form of entropy is supplied or removed via heat flows. T_{A} is the thermodynamic temperature at the point of the system boundary, where the heat flow $\dot{Q}$ appears[1].
- **Entropy transport coupled to mass flows** $\dot{m} \cdot s$: This form of entropy is supplied or removed via mass flows.

As an implication of the second law of Thermodynamics, heat can only flow in the direction of decreasing temperature. The different effects of heat transfer are explained in the following.

2.1.3 Heat Transfer

Heat transfer can take place in form of conduction, convection, or radiation. In most cases, all three effects appear simultaneously. This chapter gives a brief introduction to the fundamentals of heat transfer, based on (Incropera 2007, Frahm et al. 2022b).

2.1.3.1 Conduction

Heat conduction describes the flow of heat between two nodes, which occurs spontaneously from hot to cold without any external driving energy. The one-dimensional stationary heat transfer across a plane $\dot{Q}_{\mathrm{a}\rightarrow\mathrm{b}_{\mathrm{conduct}}}$ from the temperature T_{a} to T_{b} can be simplified by Fourier's law in Eq. (2.4),

$$\dot{Q}_{\mathrm{a}\rightarrow\mathrm{b}_{\mathrm{conduct}}} = \frac{kA}{L}(T_{\mathrm{a}} - T_{\mathrm{b}}) \tag{2.4}$$

where k describes the thermal conductivity, A the plane area and L the plane thickness (Incropera 2007).

[1] The thermodynamic temperature is a universal, not negative temperature. As the heat flow is often transmitted over a non-constant temperature interval, T_{A} is usually simplified to a mean temperature at which the heat flow occurs.

2.1.3.2 Convection

Convective heat transfer is a superposition of heat conduction and the movement of fluids. In contrast to pure heat conduction, convective heat transfer is also driven by the velocity of the fluid, e.g. wind, and is taken into account by the heat transfer coefficient,

$$\dot{Q}_{\mathrm{a}\rightarrow\mathrm{b}_{\mathrm{convect}}} = hA(T_{\mathrm{a}} - T_{\mathrm{b}}) \tag{2.5}$$

with the heat transfer coefficient h and the contact area A (Incropera 2007).

2.1.3.3 Radiation

Thermal radiation is a form of heat transfer in which a surface transfers energy in all directions at the speed of light. The radiation is caused by the thermal movement of particles. In contrast to heat conduction and convection, heat transfer depends on temperatures in the fourth order. Equation. (2.6) represents the radiative heat transfer over a distance between two gray bodies in line of sight,

$$\dot{Q}_{\mathrm{a}\rightarrow\mathrm{b}_{\mathrm{radiate}}} = \frac{A_{\mathrm{a}}\sigma(T_{\mathrm{a}}^4 - T_{\mathrm{b}}^4)}{\frac{1}{F} + \frac{1-\epsilon_{\mathrm{a}}}{\epsilon_{\mathrm{a}}} + \frac{A_{\mathrm{a}}(1-\epsilon_{\mathrm{b}})}{A_{\mathrm{b}}\epsilon_{\mathrm{b}}}} \tag{2.6}$$

with the parameters: emissivity of the surfaces ϵ, surfaces A, Stefan-Boltzmann constant σ, and visibility factor between two surfaces F (Incropera 2007).

2.1.3.4 Combining Heat Transfer Effects

In most cases, all three heat transfer effects - conduction, convection and radiation - occur simultaneously. These three effects are all driven by a temperature difference. A linearization and superposition of the equations (2.4), (2.5) and (2.6) leads to a simplified, equivalent heat flow $\dot{Q}_{\mathrm{a}\rightarrow\mathrm{b}}$, which depends only on a temperature difference and a single resistance. This equivalent heat flow is shown in equation (2.7),

$$\dot{Q}_{\mathrm{a}\rightarrow\mathrm{b}} = \frac{T_{\mathrm{a}} - T_{\mathrm{b}}}{R_{\mathrm{a,b}}} \tag{2.7}$$

where $R_{\mathrm{a,b}}$ is the thermal resistance between the nodes a and b. This approach for the thermal resistance is an exact approximation for convection and conduction (e.g. in the case of convection only: $R_{\mathrm{a,b}} = \frac{1}{hA}$, see Eq. (2.5)), while for radiation it is only a linear approximation (Frahm et al. 2022b).

2.2 Thermal Building Models

2.2.1 Physical and Data-driven Modeling Approaches

Thermal building modeling approaches are typically divided into white-box, black-box and grey-box approaches (Drgoňa et al. 2020). An overview of these three main modelings paradigms is summarized in the following.

- **White-box**: White-box models are complex, high-order physical models. These models are founded upon the principles of thermodynamics and heat transfer, necessitating a comprehensive understanding of the thermal system and its associated parameters. Typically, these models are simulated in specialized software, such as EnergyPlus (Geiger et al. 2018).
- **Black-box**: Black-box models are purely data-driven approaches that use only input-output data to identify system behavior. As a result, these models require no explicit physical context or a defined physical structure (Ferkl and Jan Široký 2010).
- **Grey-box**: The grey-box approach combines the strengths of both white-box and black-box methodologies. It employs physical knowledge of the system to determine a general model structure and identifies model parameters based on data. The resulting model structure is of low order and often utilizes the resistor–capacitor (RC) analogy (see Sec. 2.2.2) (Frahm et al. 2022b).

2.2.2 RC Analogy

RC building models are often applied as simplified, low-order physical models (Crabb et al. 1987). The RC models derive from the laws of Thermodynamics and heat transfer, using the three main simplifications (Frahm et al. 2022b):

- Work flows $\dot{W}$ and enthalpy flows $\dot{H}$ are neglected or represented by equivalent heat flows $\dot{Q}$.
- The state of the internal energy depends only on the state of the temperature T and the constant capacitance C: $\frac{\mathrm{d}U}{\mathrm{d}t} = C\frac{\mathrm{d}T}{\mathrm{d}t}$.
- Heat transfer effects between temperature nodes $\dot{Q}_{\mathrm{a}\rightarrow\mathrm{b}}$ are represented by an equivalent resistor that combines conduction, convection, and radiation (see Sec. 2.1.3.4).

In summary, the RC model can be designed with Eq. (2.8) and (2.9).

$$C\frac{\mathrm{d}T(t)}{\mathrm{d}t} = \sum_i \dot{Q}_i(t) \tag{2.8}$$

$$\dot{Q}_{\mathrm{a}\rightarrow\mathrm{b}} = \frac{\Delta T_{\mathrm{a,b}}}{R_{\mathrm{a,b}}} \tag{2.9}$$

Consequently, the thermal behavior of a building is determined by resistors and capacitors. Thermal resistors R describe possible paths of heat flows and their rate of transferred energy. Thermal capacitors C quantify the thermal capacity of thermal elements. In a building, these elements can be, for example, the air, the walls or the furniture.

To capture the thermal state of an entire building with a model, the building is simplified into a discrete number of states. This approach is also known as the lumped capacitance method. The number of states and the model structure must be sufficiently defined so that the thermal states can adequately describe the heat transfer effects associated with the thermal elements (Wen and Mishra 2018). The required complexity of the model structure depends on the use case of the applied model.

2.2.3 Thermal Zones and Model Structures

A thermal building model consists of one or more thermal zones. A model with only one zone is called "single-zone" model, while a "multi-zone" model is defined by two or more thermal zones. For multi-zone thermal building model (MZTBM)s, different types exist that differ by the degree of interaction between the individual zones.

According to the literature (Shin and Haberl 2019), a thermal zone is a space or collection of spaces with sufficiently similar space conditioning requirements. In addition, each thermal zone is operated by at least one temperature controller.

For MZTBMs, (Atam and Helsen 2016) defined two main approaches, decentral and central.

- **Decentral**: The decentral approach defines a single-zone model for each room independently. Interactions between adjacent thermal zones are not considered explicitly. As a result, parameters are decoupled between rooms because the models have no shared parameters. The parameter identification can be performed separately for each room.
- **Central**: The central approach explicitly considers the interaction between thermal zones. The interaction between zones is characterized by resistors as connections between zones.

For central MZTBMs, different degrees of centralization exist. For a low degree of centralization, the rooms are only connected to shared floor elements that connect the rooms indirectly. For a higher degree of centralization, adjacent rooms are also directly connected by resistors with each other.

In summary, decentral and central models differ in the way they consider interaction between thermal zones. While central models explicitly add resistors to represent heat flows between adjacent thermal zones, decentral models consider each thermal zone independently.

2.3 Occupants in Buildings

"A range of studies have confirmed that occupants significantly affect the energy requirements of a building through varying use patterns and energy-related behaviours" (Naylor et al. 2018). These energy-related behaviors are summarized by Carlucci et al. (2020) and can be grouped as follows:

- occupant presence and activity,
- thermostat and clothing adjustment,
- appliance use,
- electric lighting,
- windows operation,
- solar shading operation.

As defined by Carlucci et al. (2020), the occupant behavior (OB) can be grouped into occupants' presence and occupants' actions. In context of the model predictive heating control, the occupant's presence is relevant since the indoor thermal requirements differ between occupied and absence hours (Dong et al. 2022). The occupants' actions are of the relevance for the heating MPC because these actions cause either (i) heat flows (e.g. in case of sun shading adjustments or window openings) or (ii) internal heat gains (e.g. appliance use or lighting). Lastly, the OB in terms of thermostat adjustments is of direct relevance for the heating MPC because of its impact on the required heating flows.

Occupant Presence

According to Carlucci et al. (2020), occupant presence can be separated into detection, estimation, and prediction. The detection of presence and absence is defined by binary variables, estimation counts the number of occupants and prediction is a forecast of the future. Presence can be measured directly by sensor-based technologies or indirectly by probabilistic methods based on existing data (e.g. smart meters, power load, CO_2 concentration, temperature, humidity, light, wireless LAN, or cameras) (Carlucci et al. 2020).

Thermal Comfort as Driver for Thermostat Adjustments

Set point adjustments is a consequence of the thermal requirements that are driven by occupants' need for thermal comfort (Carlucci et al. 2020). Several thermal variables exist that have an impact on the thermal sensation and resulting required temperature set points. Typically, occupants adjust their thermal sensations by choosing an appropriate clothing level and air temperature (e.g. by heating, cooling, or operation of windows and shadings). In addition to these two variables, there are four others: mean radiant temperature, air velocity, relative humidity, and metabolic rate (Tartarini et al. 2020)[2]. These six variables are illustrated in Fig. 2.1 and explained in the following.

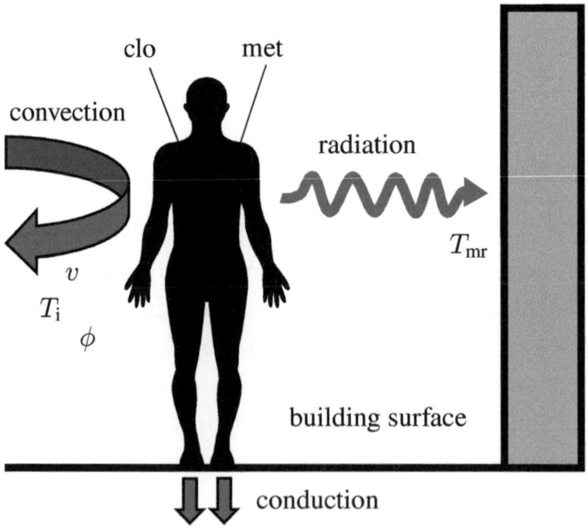

Figure 2.1: Thermal environment and heat transfer effects for thermal comfort in buildings (Frahm et al. 2022b), inspired by (Sonta 2021)

[2] CBE Thermal Comfort Tool

- **Air temperature** T_{i}: Air temperature inside of a thermal zone. The air temperature mainly drives the convective heat transfer between the thermal environment and the occupant.
- **Mean radiant temperature** T_{mr}: Temperature of the enclosing emitting surfaces of a thermal zone. It is involved in the radiative heat transfer with the occupant.
- **Air speed** v: Speed of the air flows inside the building. The airspeed has an influence on the convective heat transfer. Higher air flows create a larger cooling effect.
- **Relative humidity** ϕ: Relative humidity is defined as the pressure ratio between the water vapor in the air and the saturation of water vapor at the same temperature. For temperatures above 20 °C, a higher relative humidity results in a warmer thermal sensation, as the humidity hinders the body's sweating-based thermoregulation (Steadman 1979).
- **Clothing level** clo: Clothing creates thermal insulation around the human body. As a result, higher clothing levels make lower temperatures more comfortable.
- **Metabolic rate** met: Rate of energy production of the body, that depends on the level of activity. Higher activity levels cause the body to produce more heat, making lower temperatures more comfortable.

Overall, the thermal variables can be separated into environmental and personal variables. The thermal environment is characterized by the air and mean radiant temperature as well as air speed and relative humidity. The personal variables are described by the clothing level and metabolic rate.

Thermal Satisfaction Several thermal satisfaction models exist that aim to determine whether thermally comfortable conditions are fulfilled. Therefore, the thermal parameters need to be rated whether they result in a sufficiently warm or cold environment. The most frequently used international standards for rating thermal parameters are "ASHRAE 55" (ANSI/ASHRAE 2017), "ISO 7730" (ISO 2005), and "EN 16798" (CEN 2019). All three standards are based on "Fanger's PMV model" (Fanger 1970).

The predicted mean vote (PMV) is a static model that is evaluated using a large group of people with a specific combination of thermal environment and personal parameters. These parameters include metabolic rate, clothing, air temperature, mean radiant temperature, air velocity, and relative humidity. In a survey, occupants report their thermal sensations on a scale from -3 (too cold) to +3 (too warm), with 0 being optimal. Fanger also developed an equation that relates PMV to predicted percentage of dissatisfied (PPD).

In addition to the PMV, other indices exist that evaluate thermal sensation. For a lower complexity, the operative temperature T_{op} is an index with fewer variables than the PMV. T_{op} depends only on indoor air temperature T_{i}, radiant temperature T_{mr}, and air velocity v, as presented in Eq. (2.10) (McIntyre 1973).

$$T_{\mathrm{op}} = \begin{cases} 0.56T_{\mathrm{i}} + 0.44T_{\mathrm{mr}} & \text{for } v \leq 0.1\,\mathrm{m\,s^{-1}} \\ \frac{0.44T_{\mathrm{mr}}+0.56\left(5-\sqrt{10v}(5-T_{\mathrm{i}})\right)}{0.44+0.56\sqrt{10v}} & \text{for } v > 0.1\,\mathrm{m\,s^{-1}} \end{cases} \tag{2.10}$$

2.4 Building Control System

The building control system is dedicated to monitoring and controlling building energy systems: heating, ventilation, and air conditioning/cooling (HVAC), lighting, shading, power systems, etc. Usually, the building control system aims to provide comfortable conditions while minimizing energy costs. Therefore, specialized control strategies can utilize available information, such as occupancy, weather, or the electricity tariff.

The three main categories for building control strategies are (a) feedback control (FBC), (b) rule-based control (RBC), and (c) model predictive control (MPC) (Wen and Mishra 2018).

2.4.1 Conventional Control Strategies

Feedback control

A feedback controller generates an input u based on the setpoint y_r and the measured output y. The standard feedback control scheme is illustrated in Fig. 2.2. In building control, typical feedback controllers are PID or two-point controllers.

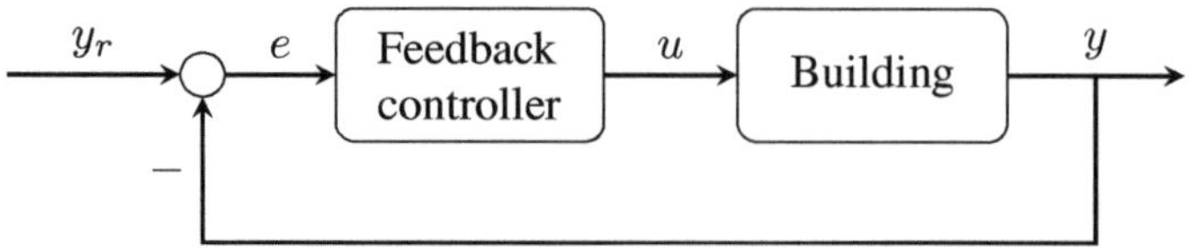

Figure 2.2: Standard feedback controller

The two-point controller switches the control input u between two states. It is also known as "hysteresis" or "bang-bang control". In thermal building control, these two states are "power on" or "power off", based on the value of the output y.

The PID controller generates a continuously varying input u. For thermal building control, the power may vary continuously, based on the measurement of the output y. However, PID cannot directly take variable regulation limitations into account, as they need to be supplemented by anti-windup measures.

Feedback controllers are well suited for clear input-output control tasks, where a single output y should be controlled by a single input u. In contrast to RBC or MPC, the feedback controller cannot consider specific rules or a cost function.

Rule-based control

RBC defines a set of rules to manipulate inputs or to schedule set points. These rules can be defined individually, e.g. by formulating case-specific “if statements”. RBC is often applied for pre-heating strategies or start/stop instructions. For example, the RBC controller price storage control (PSC) schedules heating based on thermal storage and an electricity tariff (Frahm et al. 2023).

In general, RBC ensures reliable control over a wide range of conditions. However, the control strategy can be sub-optimal or over-dimensional compared to more advanced predictive control strategies such as MPC (Wen and Mishra 2018).

2.4.2 Model Predictive Control

MPC is a constrained optimal control strategy. It calculates the optimal input trajectory over a finite horizon by minimizing a cost function. The cost function is an expression of the optimization goal, e.g. to minimize energy costs. The optimization is performed system-specifically by simulating a mathematical model of a physical system in the decision-making process. The simulation starts in each iteration with the current system states, that are obtained from measurements or observations. Future states are predicted based on the model and forecasts of time-variable parameters, such as the weather. With these predictions, the future system behavior is optimized (Drgoňa et al. 2020). The standard closed-loop MPC scheme for building control is illustrated in Fig. 2.3.

The closed-loop MPC scheme in Fig. 2.3 results from repeatedly executing an open-loop optimization. The building is affected by disturbances and controlled by an input u which results in an output y. From input and output data, an observer estimates the states $\hat{x}$. The estimation of states is required because, typically, not all states of the model are measurable. A building model represents the essential system behavior of the actual building. The building model uses the initial state x_0 from the observer and forecast of future time-variable parameter p_{tv} to calculate

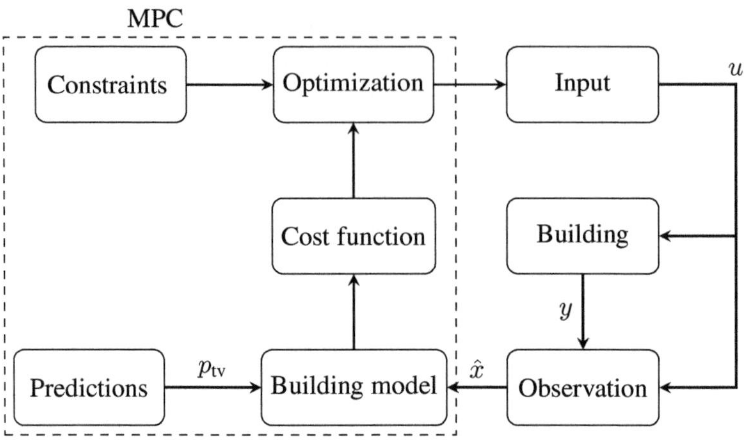

Figure 2.3: MPC closed loop formulation with an observer, based on (Drgoňa et al. 2020)

the future system behavior. Finally, the MPC controller optimizes the future system behavior as it minimizes the cost function under the consideration of constraints. Constraints can consider comfortable thermal conditions or physical limitations, e.g. from the heating system. Finally, the optimization generates the optimal input trajectory u, from which only the first value is applied to the actual building (Drgoňa et al. 2020).

2.4.3 Optimization

An optimization problem, as such present in MPC, has a form of Eq. (2.11) (Boyd and Vandenberghe 2023),

$$\begin{aligned} \min_{x} \quad & f_0(x) \\ \text{subject to:} \quad & f_i(x) \leq b_i, \quad \forall i \in [1, \dots, m] \end{aligned} \tag{2.11}$$

where $x = (x_1, \dots, x_n)^{\mathsf{T}}$ describes the optimization variable, $f_0 : \mathbf{R}^n \to \mathbf{R}$ the objective function, $f_i : \mathbf{R}^n \to \mathbf{R}$ the inequality constraint functions, and b_i bounds for the constraints. The variable x is considered optimal if it has the smallest objective value among all possible solutions that satisfy the constraints. An optimization problem is linear when the objective and constraint functions $f_0, \dots f_m$ are linear functions as in Eq. (2.12).

$$\begin{aligned} f_i(\alpha x + \beta y) = \alpha f_i(x) + \beta f_i(y), \\ \forall x, y \in \mathbf{R}^n, \ \forall \alpha, \beta \in \mathbf{R} \end{aligned} \tag{2.12}$$

An optimization problem is convex when the objective and constraint functions are convex as in Eq. (2.13).

$$
\begin{aligned}
f_i(\alpha x + \beta y) \leq \alpha f_i(x) + \beta f_i(y), \\
\forall x, y \in \mathbf{R}^n,\ \forall \alpha, \beta \in \mathbf{R},\ \alpha + \beta = 1, \alpha \geq 0, \beta \geq 0
\end{aligned}
\tag{2.13}
$$

Convexity is a more general formulation compared to linearity as the equality sign is replaced by inequality. Furthermore, the inequality needs to be fulfilled for only certain values of α and β. Although convex optimization problems cannot simply be solved analytically, there are well-developed methods to solve the convex problems efficiently while reaching the global optimum (Boyd and Vandenberghe 2023).

2.4.4 Observation

Observers estimate states that are not measured. These states are often required for feedback-based control strategies. For example, MPC needs the observation of all states as an initial starting condition (see Sec. 2.4.1c). The observation can also be used to replace sensors, which can be cost-expensive. The observer estimates the unmeasured states from measured inputs and outputs of the actual system. It is implemented as a computational algorithm, that is often based on a mathematical model and observer gain (Mohd Ali et al. 2015).

An observer can only reconstruct observable states. In the literature (Lunze 2020), complete observability describes the condition of a system, that any initial state can be reconstructed from the input and output variables. If the system is not completely observable, the observability applies to a reduced subspace of initial states inside the state space. Mathematical conditions for observability are *Observability Gramians* or *Observability Matrix*, which are both derived and explained in the literature (Mohd Ali et al. 2015).

Most observers are fundamentally based on the Luenberger observer (Luenberger 1966) or the Kalman filter (Kalman 1960). These two observers were initially designed for linear systems and were later extended to non-linear systems, e.g. the Extended Kalman filter. Furthermore, new classes of observers exist, such as AI-based (Mohd Ali et al. 2015).

Kalman Filter

The Kalman filter is an observation method for linear systems. It is explained in the following, based on (Welch and Bishop 1994).

A linear system is described by a stochastic linear differential equation system in Eq. (2.14),

$$\dot{x}(t) = Ax(t) + Bu(t) + w(t) \tag{2.14a}$$
$$y(t) = Cx(t) + v(t) \tag{2.14b}$$
$$x(0) = x_0 \tag{2.14c}$$

where x are states, u inputs, y outputs, w process noise, and v output/measurement noise. A, B, C are system matrices that contain the systems' parameters and structure. In discrete time, the system is represented by Eq. (2.15) for each step k.

$$x[k+1] = Ax[k] + Bu[k] + w[k] \tag{2.15a}$$
$$y[k] = Cx[k] + v[k] \tag{2.15b}$$
$$x[0] = x_0 \tag{2.15c}$$

The Kalman filter uses a process model and information about the statistical distribution, which is described in Eq. (2.15) and (2.16), to optimally observe the states $\hat{x}$.

$$p(w) \sim N(0, Q) \tag{2.16a}$$
$$p(v) \sim N(0, R) \tag{2.16b}$$

As presented in Eq. (2.16), process noise w and output noise v are assumed as independent of each other, white, and normally distributed. Q and R denote the covariance matrices of the process noise and output noise, respectively.

The computational implementation of the Kalman filter algorithm consists of two steps: (i) prediction and (ii) correction. These two steps are presented in Fig. 2.4.

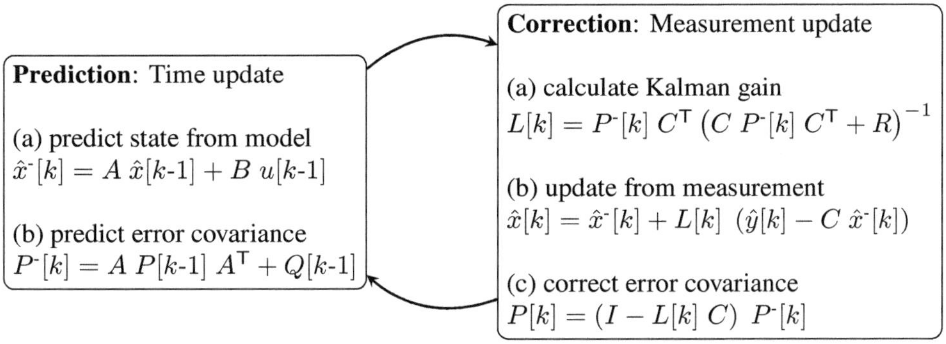

Figure 2.4: Discrete Kalman filter algorithm, based on (Welch and Bishop 1994)

In the first step, the *a priori* state $\hat{x}^{-}$ is estimated, as reflected in Eq. (2.17). The *a priori* state uses knowledge of the process prior to step k. It consists of a mean value $\hat{x}^{-}[k]$ and its error covariance $P^{-}[k]$.

$$\hat{x}^{-}[k] = A\, \hat{x}[k\text{-}1] + B\ u[k\text{-}1] \tag{2.17a}$$
$$P^{-}[k] = A\ P[k\text{-}1]\ A^{\mathsf{T}} + Q[k\text{-}1] \tag{2.17b}$$

The calculation of the *a posteriori* state is shown in Eq. (2.18). This state is also represented by a mean value $\hat{x}[k]$ and error covariance $P[k]$. The *a posteriori* state $\hat{x}[k]$ is calculated from the *a priori* estimate $\hat{x}^{-}[k]$ and a weighted difference between output measurement $\hat{y}[k]$ and its prediction $\hat{y}^{-}[k] = C\ \hat{x}^{-}[k]$. This difference is also called *residual*. The Kalman gain $L[k]$ minimizes the *a posteriori* error covariance $P[k]$.

$$L[k] = P^{-}[k]\ C^{\mathsf{T}} \left(C\ P^{-}[k]\ C^{\mathsf{T}} + R\right)^{-1} \tag{2.18a}$$
$$\hat{x}[k] = \hat{x}^{-}[k] + L[k]\,(\hat{y}[k] - C\ \hat{x}^{-}[k]) \tag{2.18b}$$
$$P[k] = (I - L[k]\ C)\ P^{-}[k] \tag{2.18c}$$

The impact of the process noise Q and the output noise covariance matrix R on the Kalman gain L can be evaluated in Eq. (2.19) and (2.20). When the measurement error covariance matrix R reaches 0, the Kalman gain L weights the residual more heavily. With smaller process noise Q, the *a priori* error covariance matrix P^{-} decreases (see Eq. (2.17)). In that case, the residual is weighted less heavily in the Kalman gain L. As a result, the difference between the *a posteriori* $\hat{x}$ and *a priori* state $\hat{x}^{-}$ decreases.

$$\lim_{R[k]\to 0} L[k] = C^{-1}[k] \tag{2.19}$$

$$\lim_{P^{-}[k]\to 0} L[k] = 0 \tag{2.20}$$

2.4.5 Subspace Identification Methods

Subspace identification methods (SIM) estimate linear time-invariant (LTI) state-space models of multiple input multiple output (MIMO) dynamic systems that are subject to input, output, and state noise. In contrast to conventional (non-linear) optimization-based identification, SIM formulate an intermediate step before identifying system parameters. First, a matrix is derived that is an essential subspace of the system to be identified. Such essential subspaces are the extended observability matrix or a sequence of Kalman states. Then, from that subspace, the system matrices are derived. Both derivations use efficient and well-developed methods from linear algebra, such as QR or singular value decomposition, and/or convex optimization methods (Qin 2006, Verhaegen 2015).

3 A New Framework for Occupant-Oriented Demand Response

As a novelty, this thesis proposes a framework for MPC-based DR in a residential building with multiple zones and distributed energy sources while considering individual occupancy per room. The concept of incorporating OB into advanced building control was first presented in (Frahm 2021) and (Dong et al. 2022). The occupant-oriented DR concept was then implemented and evaluated in (Frahm et al. 2022d) and extended to multi-zone control in (Frahm et al. 2023). Finally, the concept was extended to distributed energy sources in (Langner et al. 2024b) and (Langner et al. 2024a). This thesis brings all these aspects together in one unifying framework.

The chapter is organized as follows. Sec. 3.1 introduces components of the proposed framework while Sec. 3.2-3.4 focus on detailed descriptions of model components and constraints. Sec. 3.2 presents the multi-zone thermal building model. Next, Sec. 3.3 details the building energy management system (BEMS), including heat pump, PV, and battery, followed by Sec. 3.4 which outlines the OTS model.

3.1 Framework Introduction

Fig. 3.1 showcases the framework, illustrating OB, electricity tariff, weather predictions, time-variable limitations, and a building model.

The framework is based on MPC and implemented in the Python toolbox "do-mpc" (Lucia et al. 2017). It uses four categories for variables: time-variable parameters p_{tv}, states x, control inputs u, and control outputs[1] y. These variables can be expressed as time-continuous variables (t) or as time-discrete variables $[k]$ and these four categories are defined as follows:

[1] The control outputs are typically available as measurements.

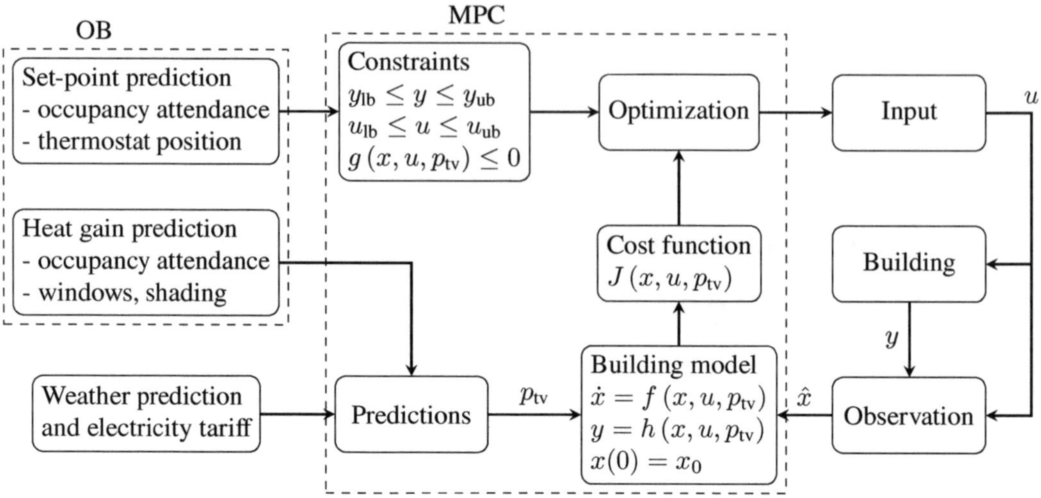

Figure 3.1: A new framework for occupant-oriented demand response, first presented in (Frahm 2021) and applied in (Dong et al. 2022) as well as (Frahm et al. 2022d)

- **Time-variable parameters** p_{tv}: Weather data (ambient temperature T_{a} and solar radiation $\dot{q}_{\text{s}}$), dynamic electricity tariff Ω_{buy}, and heat gains $\dot{Q}_{\text{g}_j}$.
- **States** x: States of the temperatures in each thermal zone j ($j = 1, \ldots, n$) (indoor air temperature T_{i_j} and heat accumulating medium T_{m_j}) as well as battery state E_{batt}.
- **Control inputs** u: Power for charging $\dot{W}_{\text{ch}}$ and discharging of the battery $\dot{W}_{\text{dc}}$ as well as heat flows into each thermal zone $\dot{Q}_{\text{h}_j}$.
- **Control outputs** y: Indoor air temperatures in each thermal zone T_{i_j}, and battery state of energy (SoE).

Mathematically, the variables of p_{tv}, x, u, and y are formulated in Eq. (3.1)-(3.4).

$$p_{\text{tv}} = \left(T_{\text{a}}\ \dot{q}_{\text{s}}\ \Omega_{\text{buy}}\ \dot{Q}_{\text{g}_1}\ \ldots\ \dot{Q}_{\text{g}_n}\right)^{\mathsf{T}} \tag{3.1}$$

$$x = \left(E_{\text{batt}}\ T_{\text{i}_1}\ T_{\text{m}_1}\ \ldots\ T_{\text{i}_n}\ T_{\text{m}_n}\right)^{\mathsf{T}} \tag{3.2}$$

$$u = \left(\dot{W}_{\text{ch}}\ \dot{W}_{\text{dc}}\ \dot{Q}_{\text{h}_1}\ \ldots\ \dot{Q}_{\text{h}_n}\right)^{\mathsf{T}} \tag{3.3}$$

$$y = \left(\text{SoE}\ T_{\text{i}_1}\ \ldots\ T_{\text{i}_n}\right)^{\mathsf{T}} \tag{3.4}$$

Overall, the framework consists of six major parts: (i) building model, (ii) OB, (iii) constraints, (iv) predictions, (v) cost function, and (vi) optimization (see Fig. 3.1) The six parts are presented in the following. The model and constraints are explained more in detail in Sec. 3.2-3.4.

3.1.1 Building Model

The building model is used to make future predictions for the decision-making process of the optimal input trajectory. The building model of this framework consists of four parts: (a) multi-zone thermal building model, (b) heat pump model, (c) PV model, and (d) battery model.

- **Multi-zone thermal building model**: The multi-zone thermal building model defines a dynamic relationship between weather data, temperature states, and heat flows in each thermal zone.
- **Heat pump model**: The heat pump model calculates the heat flows generated from the electric power consumption of the heat pump. The efficiency of energy conversion is represented by the coefficient of performance (COP).
- **PV model**: The PV model represents the generation of self-produced energy from solar energy. The produced energy can be used by the heat pump, it can be stored in the battery, or sold to the grid.
- **Battery model**: The battery model is used for the stored internal chemical energy. The battery can be both, charged and discharged, under losses.

Overall, the four model components of the building model determine the most significant energy flows. The four models are summarized to the overall building model, using the general control notation in Eq. (3.5),

$$\dot{x}(t) = f\left(x(t), u(t), p_{\text{tv}}(t)\right) \tag{3.5a}$$

$$y(t) = h\left(x(t), u(t), p_{\text{tv}}(t)\right) \tag{3.5b}$$

$$x(0) = x_0 \tag{3.5c}$$

where $\dot{x}$ is the derivative of the states x, x_0 the initial states, $f(\cdot)$ a vector function equal to the state derivatives $\dot{x}$, and $h(\cdot)$ a vector function equal to the control outputs y. The equations of building model, resulting in $f(\cdot)$ and $h(\cdot)$, will be described in Sec. 3.2 and 3.3.

For the computational implementation, the general control notation is reformulated with discrete time steps k.

$$x[k+1] = f\left(x[k], u[k], p_{\mathrm{tv}}[k]\right) \tag{3.6a}$$

$$y[k] = h\left(x[k], u[k], p_{\mathrm{tv}}[k]\right) \tag{3.6b}$$

$$x[0] = x_0 \tag{3.6c}$$

3.1.2 Occupant Behavior

As presented in (Dong et al. 2022), the energy-related OB prediction can be classified into (i) temperature set-points and (ii) behavior that impacts internal heat gains. These predictions can be considered in the optimization the address specific occupants' preferences and to reduce energy costs.

Set-point prediction The required set-points can be predicted based on the occupants' presence and the position of thermostats. During occupancy hours, thermally satisfying indoor conditions are required. On the contrary, when no occupants are present, the requirement for space heating can be reduced. In addition to estimated occupancy, the temperature preferences of occupants can be identified from thermostat positions. Based on the past use of thermostats, the individual thermal preferences of occupants and specific thermal zones can be used as an indicator for preferred set-points. Finally, the resulting setpoints can be used for shaping the constraints of the MPC, e.g. as time-variable temperature constraints.

Internal heat gain prediction Heat flows can be predicted from the occupants' presence as well as the operation of windows and shadings. Occupants cause heat gains from metabolism and the use of appliances (e.g. computers or white appliances). The operation of windows generates an energy transport coupled to mass flows (see Eq. (2.2)). An opened window transports ambient air into the indoor environment. The operation of shadings determines the heat gains from the solar radiation through the windows into the indoor environment. Closed shadings reduce the heat gains generated by solar radiation. Finally, the resulting heat gains can be used as predicted time-variable parameters and are integrated in the building model.

Summary In summary, the consideration of OB can shape temperature constraints and predict heat gains. Including these time-variable constraints and heat gains into the framework can provide thermally satisfying indoor conditions, while reducing energy costs. Most importantly, when no occupants are present, heating can be scheduled cost-efficiently.

3.1.3 Constraints

Constraints are separated into three categories: output constraints, input constraints, and general constraints. All types of constraints can be formulated as time-constant or time-variable and as hard or soft constraints. Input and output constraints limit the inputs and outputs between lower bounds (lb) and upper bounds (ub). The general constraints define various inequality functions between all types of variables. The three types of constraints are explained in detail in the following.

- **Output constraints** $y_{\text{lb}} \leq y \leq y_{\text{ub}}$: The output constraints ensure that the control outputs remain in an adequate range. For example, the temperature in a thermal zone j should remain in a permitted (time-variable) temperature range $[T_{\text{lb}_j}(t), T_{\text{ub}_j}(t)]$.
- **Input constraints** $u_{\text{lb}} \leq u \leq u_{\text{ub}}$: The input constraints limit the inputs to physically feasible values. For example, the electrical power $\dot{W}_{\text{el}}$ should remain between 0 and a maximal value.
- **General constraints** $g(x, u, p_{\text{tv}}) \leq 0$: The general constraints can define various functions between states x, inputs u, and time-variable parameters p_{tv}. The inequality sign can be inverted from $\leq$ to $\geq$ by multiplying $g(\cdot)$ with -1.

Overall, the constraints can define all possible inequality functions for all types of variables, including inputs, outputs, and states.

3.1.4 Predictions

The predictions collect data from various sources (weather, electricity grid, OB) and generate time-variable parameters p_{tv}. These parameters are used by the building model for the predictions of the overall system. This framework uses predictions of the ambient temperature T_{a}, solar radiation $\dot{q}_{\text{s}}$, dynamic electricity price Ω_{buy}, and total internal heat gains $\dot{Q}_{\text{g}_j}$.

The predictions are used to reduce energy costs over the predictive horizon of the optimization. The energy use may be reduced by accounting for future heat gains from the weather (T_{a} and

$\dot{q}_s$), but also from internal heat sources ($\dot{Q}_{g_j}$). The ambient temperature T_a is also used to optimally schedule heat flow generation of the heat pump. The efficiency of the heat pump's power conversion, the COP, is a function of the ambient temperature T_a (see Sec. 3.3.2). Finally, the electricity cost of the heat pump power consumption can be further optimized by accounting for the variability of the electricity price Ω_{buy}.

3.1.5 Cost Function

A cost function is used to define the optimization goal while considering the building model. The general term of the cost function $J\left(x, u, p_{\text{tv}}\right)$ is presented in Eq. (3.7).

$$J\left(x, u, p_{\text{tv}}\right) = \underbrace{m(x[N+1])}_{\text{Meyer term}} + \sum_{k=0}^{N} \left(\underbrace{l(x[k], u[k], p_{\text{tv}}[k])}_{\text{Lagrange term}} + \underbrace{\Delta u[k]^{\mathsf{T}} R \Delta u[k]}_{\text{R-term}} \right) \tag{3.7}$$

The cost function consists of three parts, the Meyer term, the Lagrange term, and the R-term.

- **Meyer term** $m(\cdot)$: The Meyer terms defines the cost of the terminal state $x[N+1]$. This term is used when a specific state should be reached in the final step $N+1$.
- **Lagrange term** $l(\cdot)$: The Lagrange term describes the costs of each stage k ($k = 0 \ldots N$). It defines the main optimization goal over the entire period of optimization.
- **R-term**: The R-term accounts for changes in the inputs. The R-term prevents an oscillating behavior resulting from quick changes in the inputs. It can also account for physical limitations in input changes.

In this thesis, the Lagrange term defines the operational costs, resulting from the power bought or sold and the electricity tariffs.

$$l(x[k], u[k], p_{\text{tv}}[k]) = \dot{W}_{\text{buy}}[k] \cdot \Omega_{\text{buy}}[k] - \dot{W}_{\text{sell}}[k] \cdot \Omega_{\text{sell}}[k] \tag{3.8}$$

The optimization of this cost function finds the optimal strategy for buying, selling, and consuming/storing energy, as described in the following.

3.1.6 Optimization

The cost function $J(x, u, p_{\mathrm{tv}})$ is used to solve a constrained optimization problem in Eq. (3.9), as presented in (Frahm et al. 2023). Thanks to the output constraints, a certain terminal state is not required. As a result, the Meyer term $m(\cdot)$ from Eq. (3.7) can be neglected.

$$\min_{\mathbf{x}[0:N+1],\mathbf{u}[0:N]} \sum_{k=0}^{N} \left(l(x[k], u[k], p_{\mathrm{tv}}[k]) + \Delta u[k]^{\mathsf{T}} R \Delta u[k]\right) \tag{3.9a}$$

subject to $\forall k \in [0, N]$:

$$x[0] = x_0 \tag{3.9b}$$

$$x[k{+}1] = f\left(x[k], u[k], p_{\mathrm{tv}}[k]\right) \tag{3.9c}$$

$$y[k] = h\left(x[k], u[k], p_{\mathrm{tv}}[k]\right) \tag{3.9d}$$

$$y_{\mathrm{lb}}[k] \leq y[k] \leq y_{\mathrm{ub}}[k] \tag{3.9e}$$

$$u_{\mathrm{lb}}[k] \leq u[k] \leq u_{\mathrm{ub}}[k] \tag{3.9f}$$

$$g\left(x[k], u[k], p_{\mathrm{tv}}[k]\right) \leq 0 \tag{3.9g}$$

Eq. (3.9) consists of seven subequations (a)-(g) that are explained in the following. (3.9a) minimizes a cost function $J(x, u, p_{\mathrm{tv}})$. The building model is described in (3.9b)-(3.9d) (see Sec. 3.1.1). Constraints are defined in (3.9e)-(3.9g) (see Sec. 3.1.3).

The building model consists of the MZTBM and the BEMS model, which are described in Sec. 3.2 and 3.3, respectively. These sections describe also the constraints of these models. Finally, the temperature constraints are derived in the OTS model in Sec. OTS.

3.2 Multi-Zone Thermal Building Model

The framework is applicable to various types of thermal building models, including single-zone or multi-zone thermal building models. In the following, four different model structures are explained: single-zone as well as central, decentral, and fusion multi-zone.

The multi-zone thermal building model (MZTBM) defines a dynamic relationship between the indoor air temperatures in the thermal zones T_{i_j}, heat flows $\dot{Q}_{\mathrm{h}_j}$, and weather data (T_{a} and $\dot{q}_{\mathrm{s}}$). An additional state T_{m_j} is used to represent the thermal dynamics of the heat-accumulating medium that is in heat exchange with the indoor air temperature (e.g. equivalent to walls or interiors). The relationship for each room j is presented in Fig. 3.2, using RC modeling as explained in Sec. 2.2 on page 13.

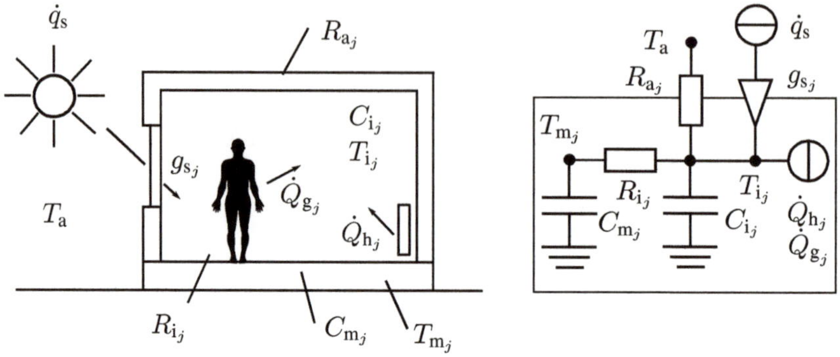

Figure 3.2: Grey-box thermal building model for each room j ($j = 1 \ldots n$), obtained from (Frahm et al. 2023), which is based on (Madsen and Holst 1995)

As a result, resistors and capacitors define the model structure. The illustrated model is based on the literature (Madsen and Holst 1995) and is explained in the following. The two temperature nodes T_{i_j} and T_{m_j} are connected by the inside resistor R_{i_j}. Furthermore, the inside air temperature node T_{i_j} is connected to the outside air temperature T_a using the outside resistor R_{a_j}. The inside air temperature T_{i_j} is also affected by heat flows from the sun $\dot{Q}_{s_j}$, internal heat gains $\dot{Q}_{g_j}$, and the heating $\dot{Q}_{h_j}$. The solar heat flow $\dot{Q}_{s_j}$ results from the global radiation $\dot{q}_s$ and the solar heat gain coefficient g_{s_j}. Internal heat gains $\dot{Q}_{g_j}$ come from occupancy attendance. Finally, each room needs to be heated by a heat flow $\dot{Q}_{h_j}$, provided by the heat pump (see Sec. 3.3.2).

Based on Fig 3.2, each room is mathematically defined by the two differential equations Eq. (3.10) and (3.11).

$$\forall j \in [1, n]: \quad C_{i_j} \frac{dT_{i_j}(t)}{dt} = \frac{T_{m_j}(t) - T_{i_j}(t)}{R_{i_j}} + \frac{T_a(t) - T_{i_j}(t)}{R_{a_j}} + g_{s_j} \dot{q}_s(t) + \dot{Q}_{h_j}(t) + \dot{Q}_{g_j}(t) \tag{3.10}$$

$$\forall j \in [1, n]: \quad C_{m_j} \frac{dT_{m_j}(t)}{dt} = \frac{T_{i_j}(t) - T_{m_j}(t)}{R_{i_j}} \tag{3.11}$$

Finally, the differential equation system can be solved to obtain a trajectory of the states $T_{i_j}(t)$ and $T_{m_j}(t)$ over time. With the solved indoor air temperature state $T_{i_j}(t)$, it can be constrained in each room j ($j = 1 \dots n$) between lower and upper bounds, as presented in Eq. (3.12). These temperature bounds may also vary over time.

$$\forall j \in [1, n] : \quad T_{\text{i,lb}_j}(t) \leq T_{\text{i}_j}(t) \leq T_{\text{i,ub}_j}(t) \tag{3.12}$$

Applying this structure in Fig. 3.2 to each room j ($j = 1 \dots n$) results in a MZTBM. A MZTBM can be established in three different forms: central, decentral, or fusion. The differences between these structures are explained in Sec. 2.2.3. For demonstration, a simple example with two rooms ($j = 1, 2$) is used.

3.2.1 Central Multi-Zone Model

The model structure of the central model is demonstrated with a two-zone model, as presented in Fig. 3.3. First, two thermal zones $j = 1, 2$ are modeled individually, applying the model structure from Fig. 3.2 and Eq. (3.10)-(3.11) to each thermal zone. Then, both zones are connected by the shared resistor $R_{1,2}$.

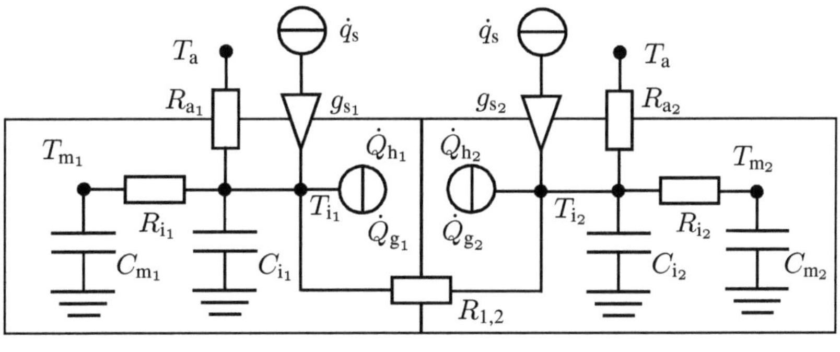

Figure 3.3: Central multi-zone thermal building model, as an extension of (Madsen and Holst 1995)

The overall system is described by four differential equations (Eq. (3.13)-(3.16)), two for each zone j ($j = 1, 2$). As a result, the individual thermal zones are explicitly connected.

$$C_{\mathrm{i}_1} \frac{\mathrm{d}T_{\mathrm{i}_1}}{\mathrm{d}t} = \frac{T_{\mathrm{m}_1} - T_{\mathrm{i}_1}}{R_{\mathrm{i}_1}} + \frac{T_{\mathrm{a}} - T_{\mathrm{i}_1}}{R_{\mathrm{a}_1}} + g_{\mathrm{s}_1}\dot{q}_{\mathrm{s}} + \dot{Q}_{\mathrm{h}_1} + \dot{Q}_{\mathrm{g}_1} + \frac{T_{\mathrm{i}_2} - T_{\mathrm{i}_1}}{R_{1,2}} \tag{3.13}$$

$$C_{\mathrm{m}_1} \frac{\mathrm{d}T_{\mathrm{m}_1}}{\mathrm{d}t} = \frac{T_{\mathrm{i}_1} - T_{\mathrm{m}_1}}{R_{\mathrm{i}_1}} \tag{3.14}$$

$$C_{\mathrm{i}_2} \frac{\mathrm{d}T_{\mathrm{i}_2}}{\mathrm{d}t} = \frac{T_{\mathrm{m}_2} - T_{\mathrm{i}_2}}{R_{\mathrm{i}_2}} + \frac{T_{\mathrm{a}} - T_{\mathrm{i}_2}}{R_{\mathrm{a}_2}} + g_{\mathrm{s}_2}\dot{q}_{\mathrm{s}} + \dot{Q}_{\mathrm{h}_2} + \dot{Q}_{\mathrm{g}_2} + \frac{T_{\mathrm{i}_1} - T_{\mathrm{i}_2}}{R_{1,2}} \tag{3.15}$$

$$C_{\mathrm{m}_2} \frac{\mathrm{d}T_{\mathrm{m}_2}}{\mathrm{d}t} = \frac{T_{\mathrm{i}_2} - T_{\mathrm{m}_2}}{R_{\mathrm{i}_2}} \tag{3.16}$$

3.2.2 Decentral Multi-Zone Model

It is also possible to consider each thermal zone individually and neglect the explicit connection over shared resistors. This results in a decentral MZTBM, as presented in Fig. 3.4.

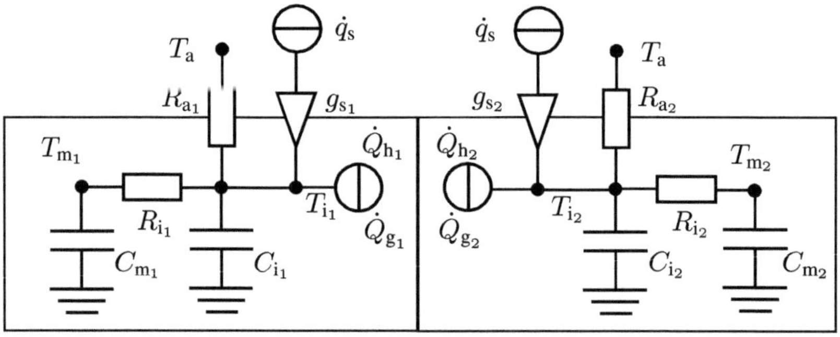

Figure 3.4: Decentral multi-zone thermal building model, as an extension of (Madsen and Holst 1995)

Similarly to the central MZTBM, the overall system is described by four differential equations (Eq. (3.17)-(3.20)). However, in this case, there exists no resistor $R_{1,2}$ that explicitly connects both zones. Consequently, the decentral model structure cannot account for heat exchange between zones.

$$C_{\mathrm{i}_1}\frac{\mathrm{d}T_{\mathrm{i}_1}}{\mathrm{d}t} = \frac{T_{\mathrm{m}_1}-T_{\mathrm{i}_1}}{R_{\mathrm{i}_1}} + \frac{T_{\mathrm{a}}-T_{\mathrm{i}_1}}{R_{\mathrm{a}_1}} + g_{\mathrm{s}1}\dot{q}_{\mathrm{s}} + \dot{Q}_{\mathrm{h}_1} + \dot{Q}_{\mathrm{g}1} \tag{3.17}$$

$$C_{\mathrm{m}_1}\frac{\mathrm{d}T_{\mathrm{m}_1}}{\mathrm{d}t} = \frac{T_{\mathrm{i}_1}-T_{\mathrm{m}_1}}{R_{\mathrm{i}_1}} \tag{3.18}$$

$$C_{\mathrm{i}_2}\frac{\mathrm{d}T_{\mathrm{i}_2}}{\mathrm{d}t} = \frac{T_{\mathrm{m}_2}-T_{\mathrm{i}_2}}{R_{\mathrm{i}_2}} + \frac{T_{\mathrm{a}}-T_{\mathrm{i}_2}}{R_{\mathrm{a}_2}} + g_{\mathrm{s}2}\dot{q}_{\mathrm{s}} + \dot{Q}_{\mathrm{h}_2} + \dot{Q}_{\mathrm{g}2} \tag{3.19}$$

$$C_{\mathrm{m}_2}\frac{\mathrm{d}T_{\mathrm{m}_2}}{\mathrm{d}t} = \frac{T_{\mathrm{i}_2}-T_{\mathrm{m}_2}}{R_{\mathrm{i}_2}} \tag{3.20}$$

3.2.3 Single-Zone Model

In a single zone model, the entire building is modeled as one thermal zone as presented in Fig. 3.5.

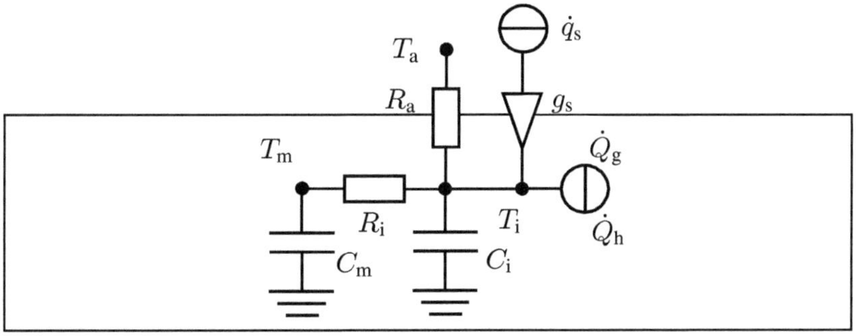

Figure 3.5: Single-zone thermal building model, based on (Madsen and Holst 1995)

As a result, the entire system is described by just one thermal zone. All states and parameters are equal in all zones, e.g. for the indoor air temperature: $T_{\mathrm{i}_1} = T_{\mathrm{i}_2} = T_{\mathrm{i}}$. Therefore, this model can be described by only the following two differential equations (Eq. (3.21) and (3.22)).

$$C_{\mathrm{i}}\frac{\mathrm{d}T_{\mathrm{i}}}{\mathrm{d}t} = \frac{T_{\mathrm{m}}-T_{\mathrm{i}}}{R_{\mathrm{i}}} + \frac{T_{\mathrm{a}}-T_{\mathrm{i}}}{R_{\mathrm{a}}} + g_{\mathrm{s}}\dot{q}_{\mathrm{s}} + \dot{Q}_{\mathrm{h}} + \dot{Q}_{\mathrm{g}} \tag{3.21}$$

$$C_{\mathrm{m}}\frac{\mathrm{d}T_{\mathrm{m}}}{\mathrm{d}t} = \frac{T_{\mathrm{i}}-T_{\mathrm{m}}}{R_{\mathrm{i}}} \tag{3.22}$$

3.2.4 Fusion Multi-Zone Model

The fusion model combines the decentral MZTBM with the single-zone model, as shown in Fig. 3.6. Similarly to the decentral MZTBM, each zone is modeled individually without explicit connections. While the decentral model uses individual thermal parameters for each zone, the parameters of the fusion model are all equal, similarly to the single-zone model. In contrast to the single-zone model, the thermal states of the fusion model differ in each zone.

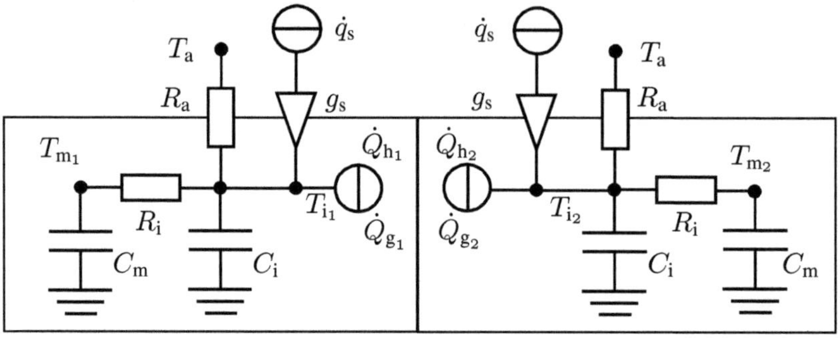

Figure 3.6: Fusion multi-zone thermal building model, as an extension of (Madsen and Holst 1995)

Similarly to the decentral MZTBM, the overall system is described by four differential equations (Eq. (3.23)-(3.26)).

$$C_\mathrm{i}\frac{\mathrm{d}T_{\mathrm{i}_1}}{\mathrm{d}t} = \frac{T_{\mathrm{m}_1} - T_{\mathrm{i}_1}}{R_\mathrm{i}} + \frac{T_\mathrm{a} - T_{\mathrm{i}_1}}{R_\mathrm{a}} + g_\mathrm{s}\dot{q}_\mathrm{s} + \dot{Q}_{\mathrm{h}_1} + \dot{Q}_{\mathrm{g}_1} \tag{3.23}$$

$$C_\mathrm{m}\frac{\mathrm{d}T_{\mathrm{m}_1}}{\mathrm{d}t} = \frac{T_{\mathrm{i}_1} - T_{\mathrm{m}_1}}{R_\mathrm{i}} \tag{3.24}$$

$$C_\mathrm{i}\frac{\mathrm{d}T_{\mathrm{i}_2}}{\mathrm{d}t} = \frac{T_{\mathrm{m}_2} - T_{\mathrm{i}_2}}{R_\mathrm{i}} + \frac{T_\mathrm{a} - T_{\mathrm{i}_2}}{R_\mathrm{a}} + g_\mathrm{s}\dot{q}_\mathrm{s} + \dot{Q}_{\mathrm{h}_2} + \dot{Q}_{\mathrm{g}_2} \tag{3.25}$$

$$C_\mathrm{m}\frac{\mathrm{d}T_{\mathrm{m}_2}}{\mathrm{d}t} = \frac{T_{\mathrm{i}_2} - T_{\mathrm{m}_2}}{R_\mathrm{i}} \tag{3.26}$$

3.3 Building Energy System Model

3.3.1 Power Balance

The building energy system model fulfills the energy balance, based on the first law of Thermodynamics (see Eq. (2.2)). The energy balance is presented in Eq. (3.27) in energy derivative form.

$$\dot{W}_{\mathrm{grid}}(t) = \dot{W}_{\mathrm{h}}(t) - \dot{W}_{\mathrm{PV}}(t) + \dot{W}_{\mathrm{ch}}(t) - \dot{W}_{\mathrm{dc}}(t) \tag{3.27}$$

The interaction between the building energy system and the electricity grid is described by the grid power $\dot{W}_{\mathrm{grid}}$. The grid can provide electrical power for heating $\dot{W}_{\mathrm{h}}$ and battery charging $\dot{W}_{\mathrm{ch}}$. It can also remove power from, e.g. excessive power from PV $\dot{W}_{\mathrm{PV}}$ or battery discharging $\dot{W}_{\mathrm{ch}}$.

The grid power $\dot{W}_{\mathrm{grid}}$ is separated in buying $\dot{W}_{\mathrm{buy}}$ and selling power $\dot{W}_{\mathrm{sell}}$, depending on the sign, as presented in Eq. (3.28) and (3.29), respectively. The bought and sold power are both part of the Lagrange term in Eq. (3.8) on page 30.

$$\dot{W}_{\mathrm{buy}}(t) = \max\left\{\dot{W}_{\mathrm{grid}}(t), 0\right\} \tag{3.28}$$

$$\dot{W}_{\mathrm{sell}}(t) = \min\left\{\dot{W}_{\mathrm{grid}}(t), 0\right\} \tag{3.29}$$

3.3.2 Heat Pump Model

The heat pump is the major energy consumer, providing the required heat flows for each thermal zone j. The overall heat flow produced by the heat pump is equal to the sum of heat flows for each thermal zone j, as presented in Eq. (3.30).

$$\dot{Q}_{\mathrm{h}}(t) = \sum_{j=1}^{n} \dot{Q}_{\mathrm{h}_j}(t) \tag{3.30}$$

To produce heat flows $\dot{Q}_{\mathrm{h}}$, the heat pump requires electrical power $\dot{W}_{\mathrm{h}}$. The electrical power of the heat pump depends on the COP ε_{h}, which is shown in Eq. (3.31). The COP is a time-variable parameter. In the case of an ambient air-sourced heat pump, the COP depends on the ambient air temperature T_{a}. In addition, the COP depends on the supply temperature of the heating medium to the heating distribution system (radiators or underfloor heating). Usually, with higher ambient temperatures, the COP increases, while it decreases with higher distribution temperatures.

$$\dot{W}_{\mathrm{h}}(t) = \frac{\left|\dot{Q}_{\mathrm{h}}(t)\right|}{\varepsilon_{\mathrm{h}}(t)} \tag{3.31}$$

The heat pump's power can be modulated with the modulation degree χ_{mod}. The modulation of the heat pump is presented in Eq. (3.32) and (3.33). The heat pump can be turned off with $\chi_{\mathrm{mod}} = 0$ or modulated between a lower bound $\chi_{\mathrm{mod_{lb}}}$ and 1.

$$\dot{W}_{\mathrm{h}}(t) = \chi_{\mathrm{mod}}(t) \cdot \dot{W}_{\mathrm{max}}(t) \tag{3.32}$$

$$\chi_{\mathrm{mod}} \in [0, 1] \tag{3.33}$$

Finally, the framework restricts the maximal power $\dot{W}_{\mathrm{max}}(t)$ as well as the heat flows in each thermal zone j. This ensures the physical feasibility of the heat pump.

$$0 \leq \dot{W}_{\mathrm{max}}(t) \leq \dot{W}_{\mathrm{max,ub}}(t) \tag{3.34}$$

$$\forall j \in [1, n]: \quad \dot{Q}_{\mathrm{h,lb_j}} \leq \dot{Q}_{\mathrm{h}_j}(t) \leq \dot{Q}_{\mathrm{h,ub_j}} \tag{3.35}$$

3.3.3 PV Model

The is presented in the following, based on (Al Essa 2019, Langner et al. 2024b,a). The PV model calculates the net output power of the photovoltaic system $\dot{W}_{\mathrm{PV}}$, as presented in Eq. (3.36),

$$\dot{W}_{\mathrm{PV}}(t) = \dot{W}_{\mathrm{PV_s}}(t) \cdot \eta_{\mathrm{stc}} \cdot \eta_{\mathrm{I\&C}} \cdot \eta_{\mathrm{T}}(t) \tag{3.36}$$

where $\dot{W}_{\mathrm{PV_s}}$ is the solar power, η_{stc} the efficiency of the photovoltaic module at standard testing conditions, $\eta_{\mathrm{I\&C}}$ the efficiency of inverter and controller, and η_{T} the temperature-dependent efficiency. These variables and parameters are explained and calculated in the following.

The solar power $\dot{W}_{\mathrm{PV_s}}$ depends on the surface a_{modules} and number n_{modules} of modules as well as global solar radiation $\dot{q}_{\mathrm{s}}$, as presented in Eq. (3.37).

$$\dot{W}_{\mathrm{PV_s}}(t) = a_{\mathrm{module}} \cdot n_{\mathrm{modules}} \cdot \dot{q}_{\mathrm{s}}(t) \tag{3.37}$$

The efficiency of modules under standard conditions η_{stc} as well as the efficiency of inverter and controller $\eta_{\text{I\&C}}$ can be obtained from the manufacturer's sheet (see Sec. 4.1.1). The temperature efficiency η_{T} is calculated in Eq. (3.38) (Al Essa 2019),

$$\eta_{\text{T}}(t) = 1 - \frac{T_{\text{cel}}(t) - T_{\text{stc}}}{g_{\text{T}}} \tag{3.38}$$

where T_{cel} is the cell temperature (in °C), T_{stc} the temperature of standard testing conditions (in °C), and g_{T} the temperature coefficient (in °C). The cell temperature T_{cel} depends on weather conditions and manufacturer's properties, as presented in Eq. (3.39) (Al Essa 2019).

$$T_{\text{cel}}(t) = T_{\text{a}}(t) + \dot{q}_{\text{s}}(t) \cdot g_{\text{noct}} \tag{3.39}$$

The manufacturer's property g_{noct} (in $°\text{C}\,\text{m}^2\,\text{W}^{-1}$) is calculated in Eq. (3.40) (Al Essa 2019). It results from the nominal operating cell temperature T_{noct} (in °C).

$$g_{\text{noct}} = \frac{T_{\text{noct}} - 20\,°\text{C}}{800\,\text{W}\,\text{m}^{-2}} \tag{3.40}$$

3.3.4 Battery Model

The stored (internal chemical) energy of the battery E_{batt} depends on the power used for charging $\dot{W}_{\text{ch}}$ and discharging $\dot{W}_{\text{dc}}$. Losses during charging and discharging $\dot{Q}_{\text{loss}}$ can be considered with charging and discharging efficiencies, η_{ch} and η_{dc}, respectively, as presented in Eq. (3.41). Other losses are neglected, e.g. battery standing losses.

$$\frac{\mathrm{d}E_{\text{batt}}(t)}{dt} = \dot{W}_{\text{batt,in}}(t) - \dot{W}_{\text{batt,out}}(t) - \dot{Q}_{\text{batt,loss}}(t) = \dot{W}_{\text{ch}}(t) \cdot \eta_{\text{ch}} - \frac{\dot{W}_{\text{dc}}(t)}{\eta_{\text{dc}}} \tag{3.41}$$

The battery degradation depends on storage conditions and use patterns, mainly driven by SoE, charging/discharging rate, and temperature (Woody et al. 2020). Therefore, the SoE and charging/discharging rates are constrained to reduce battery aging. In addition, the constraints ensure physical feasibility, e.g. avoiding negative values.

The power for charging and discharging is constrained in Eq. (3.42) and (3.43). The power must be positive and below an upper bound (ub).

$$0 \leq \dot{W}_{\text{ch}}(t) \leq \dot{W}_{\text{ch}_{\text{ub}}} \tag{3.42}$$

$$0 \leq \dot{W}_{\text{dc}}(t) \leq \dot{W}_{\text{dc}_{\text{ub}}} \tag{3.43}$$

The energy stored in the battery is often represented in relation to the maximal energy, as presented in Eq. (3.44). This relative value is known as the state of energy (SoE).

$$\mathrm{SoE}(t) = \frac{E_{\mathrm{batt}}(t)}{E_{\mathrm{batt}_{\mathrm{max}}}} \tag{3.44}$$

The SoE is constrained between lower and upper bounds, which is shown in Eq. (3.45).

$$\mathrm{SoE}_{\mathrm{lb}} \leq \mathrm{SoE}(t) \leq \mathrm{SoE}_{\mathrm{ub}} \tag{3.45}$$

3.4 Occupants' Thermal Satisfaction Model

The OTS model represents a temperature range schedule $[T_{\mathrm{lb}_j}(t), T_{\mathrm{ub}_j}(t)]$, that is time-variant, occupant-oriented, and room-individual. It is based on the international standards for thermal sensation and occupant behavior from Sec. 2.3. Fundamentally, the OTS model derives four different temperature ranges from PMV ranges, using the CBE Thermal Comfort Tool (Tartarini et al. 2020). These temperature ranges are applied to different, time-variable thermal demands of the occupants, e.g. room-individual attendance profiles.

The different OTS levels are presented in Tab. 3.1. The occupants' thermal satisfaction (OTS) level can be selected between different predicted mean vote (PMV) boundaries, e.g. ± 0.2 for level I or ± 0.7 for level III. The predicted percentage of dissatisfied (PPD) is a function of the PMV.

Table 3.1: OTS categories, obtained from CBE Thermal Comfort Tool (Tartarini et al. 2020) with EN-16798 and winter clothing

OTS level	PMV	PPD	T_{lb}	T_{ub}
I	± 0.2	$< 6\,\%$	22.6 °C	24.0 °C
II	± 0.5	$< 10\,\%$	21.5 °C	25.0 °C
III	± 0.7	$< 15\,\%$	20.7 °C	25.8 °C
off	-	-	16.0 °C	30.0 °C

Based on these OTS levels in Tab. 3.1, the corresponding lower T_{lb} and upper T_{ub} temperature limits are calculated with the CBE Thermal Comfort Tool (Tartarini et al. 2020). In the tool, EN-16798 standard and winter clothing are selected. In addition, the mean radiant temperature is set equal to the air temperature T_{i}. This implies the assumption that the operative temperature is close to the air temperature (see Sec. 2.3). The OTS level off is designed to maintain safe conditions in the building, e.g. to prevent mold growth. Finally, the resulting temperature limits for different levels of OTS are also presented in Tab. 3.1.

Based on the different OTS levels, three different control modes can be derived for, inspired by (Peng et al. 2018) and applied in (Frahm et al. 2023). The three modes define the space conditioning requirements in Sec. 4.1.2 on page 46, which are represented by time-variable and room-individual temperature ranges $[T_{\mathrm{lb}_j}(t), T_{\mathrm{ub}_j}(t)]$.

- **Comfort mode** OTS level I: The comfort mode aims for the highest level of OTS. The permitted temperature range is relatively small with $[22.6\,^\circ\mathrm{C},\ 24.0\,^\circ\mathrm{C}]$.
- **Eco mode** OTS level III: The eco mode extends the permitted temperature range. Compared to the comfort mode, lower minimal and higher maximal temperatures are allowed with $[20.7\,^\circ\mathrm{C},\ 25.8\,^\circ\mathrm{C}]$.
- **Standby mode** OTS level off: The standby mode aims for the lowest energy consumption with $[16.0\,^\circ\mathrm{C},\ 30.0\,^\circ\mathrm{C}]$.

The eco mode is designed to save energy compared to the comfort mode. For example, the eco mode could save energy in rooms that are less frequently occupied, e.g. bathroom or kitchen. It should also enable fast re-heating compared to the standby mode (Peng et al. 2018).

4 Evaluation of the New Framework

The performance of the proposed framework is assessed in this chapter through a comparison with other control strategies, across varying configurations, and in the presence of forecast and model uncertainties. Two different occupancy scenarios are evaluated in a new way as they differ in their complexity for multi-zone control. Furthermore, different framework configurations are investigated with regard to the additional benefit of PV and batteries when already participating in a DR program. Finally, an uncertainty analysis is conducted in multi-zone control with DR and distributed energy sources by adding noise to the model and forecasts.

Sec. 4.1 outlines the simulated environment in which different algorithms and configurations are evaluated. Sec. 4.2 presents an evaluation of the framework's general performance by comparing it with other control strategies. Sec. 4.3 assesses the framework under various configurations, including PV and battery elements. Additionally, the control performance is evaluated under the effect of forecast and model uncertainties.

4.1 Settings

The evaluation settings define the simulation environment for simulating and evaluating the different algorithms and configurations. This section is organized by building parameters, occupancy scenarios, data, uncertainties, and metrics.

4.1.1 Building Parameters

The building parameters describe the energy-related characteristics of the simulation environment. These parameters are separated into four categories: thermal building parameters, heat pump parameters, PV parameters, and battery parameters. The following parameters represent a white-box modeling approach in which they were derived from literature values, manufacturer specifications, and geometry data.

Thermal Building Parameters

The thermal building parameters describe the dynamic characteristics and the number of thermal zones of the model (see Sec. 3.2). As illustrated in Fig. 4.1, a decentral multi-zone thermal building model (MZTBM) structure with five different zones ($n = 5$) is applied (see Sec. 2.2.3). This allows the representation of various different space conditioning requirements within one environment (Frahm et al. 2023).

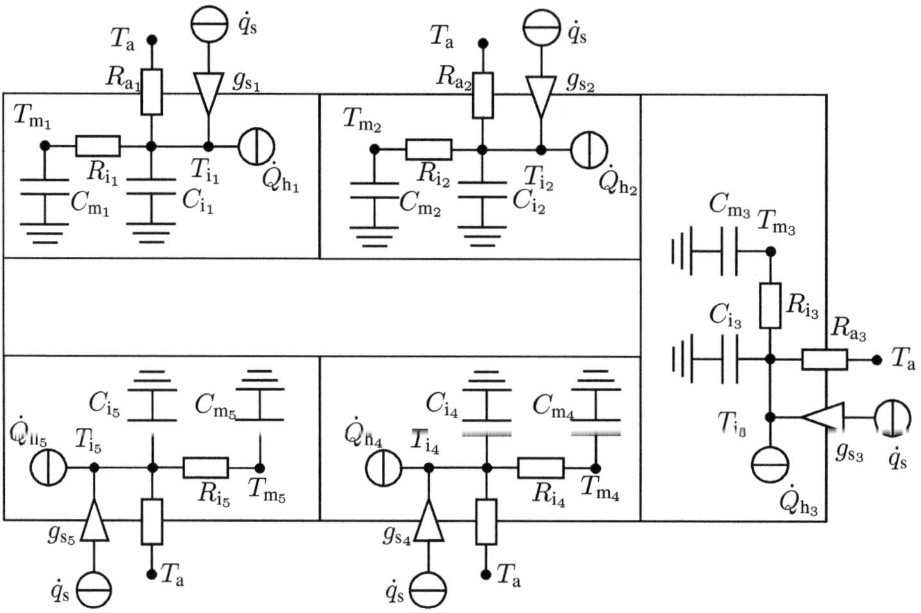

Figure 4.1: Thermal building model for evaluation, obtained from (Frahm et al. 2023)

The thermal parameters of the building model are presented in Tab. 4.1 (Frahm et al. 2023). The model parameters are obtained from (Madsen and Holst 1995) and scaled from the single-zone to the multi-zone level (Frahm et al. 2022c). To represent sufficient difference in space conditioning requirements, the number of thermal zones was selected to five (see Sec. 4.1.2). Two different parameter sets are created to represent buildings with high and low thermal capacity.

Table 4.1: Model parameters for high and low thermal capacitance (cap.), obtained from (Frahm et al. 2023)

parameters	C_i in $\mathrm{J\,K^{-1}}$	C_m in $\mathrm{J\,K^{-1}}$	R_a in $\mathrm{K\,W^{-1}}$	R_i in $\mathrm{K\,W^{-1}}$	g_s
high cap.	3407040	11482560	0.07345	0.001197	1.138
low cap.	1703520	5741280	0.07345	0.001197	1.138

The thermal parameters from Tab. 4.1 can be defined individually for each room j ($j = 1 \ldots n$), as shown in Fig. 4.1 (with $n = 5$). To represent variability in requirements for different rooms, the standard parameters from Tab. 4.1 are multiplied by different random-seed factors for each room. This random seed is once initialized and then remains constant over all simulations. The parameters may vary by up to $\pm 5\,\%$ in each room.

Heat Pump Parameters

Heat pump parameters are available in Tab. 4.2. As described in (Frahm et al. 2023), the air-source heat pump parameters can be found in the manufacturer's fact sheet[1]. The manufacturer's technical sheet provides the COP ε_h and the maximum electrical power consumption P_max, both as a function of the ambient temperature.

Table 4.2: Heat pump parameters for different outside air temperatures T_a, obtained from (Frahm et al. 2023)

T_a in °C	-10	-7	2	7	10	12	15	20
ε_h	1.98	2.20	2.71	3.10	3.34	3.55	3.89	4.26
P_max kW	4.20	4.39	4.83	4.62	4.40	4.41	4.00	3.32

Both parameters, ε_h and P_max, depend not only on the outside air temperature T_a but also on the heating supply temperature. The supply temperature is assumed as a constant value of 55 °C[2]. Only the variability of the ambient air temperature is considered. The influence of the ambient air temperature is represented by eight discrete values in Tab. 4.2. Between these values, linear interpolation is applied.

1 For the model "AERO SLM 3-11 HGL" (iDM Energiesysteme GmbH 2020).

2 Additional parameters for other supply temperatures can be found in the fact sheet (iDM Energiesysteme GmbH 2020) for heating and cooling.

PV Parameters

The PV parameters are outlined in Tab. 4.3, which are required to calculate the PV power generation. The generated power is determined by the solar radiation, the number and area of modules, and their efficiency. A total of 30 modules with an area of 1.685 m^2 each are installed on the experimental building. The modules are "Hanwha Q-Cells Q.Peak Duo G5" with a peak power of 320 W_p per module, resulting in a total peak power of 9.6 kW_p. The manufacturer's fact sheet provides efficiency under standard conditions η_{stc} and the nominal cell temperature T_{noct}. The remaining parameters are obtained from the literature (Langner et al. 2024a).

Table 4.3: PV parameters, obtained from (Langner et al. 2024a)

η_{stc}	$\eta_{I\&C}$	$a_{modules}$	$n_{modules}$	g_T	T_{noct}	T_{stc}
0.19	0.9	1.685 m^2	30	200 °C	45 °C	25 °C

Battery Parameters

The battery specifications are provided in Tab. 4.4. The research building employs three "LG Chem E3/DC S10E DCB-NLx" batteries, each holding 6.524 kW h, for a total of 19.57 kW h. The charging and discharging power is capped at 2.88 kW, while maintaining the SoE between 20 and 80 %. The initial and end SoE are defined as 50 %. The energy that is discharged must therefore be returned by the end of the simulation.

Table 4.4: Battery parameters

$P_{dc_{ub}}$	$P_{ch_{ub}}$	SoE_{lb}	SoE_{ub}	$E_{batt_{max}}$
2.88 kW	2.88 kW	0.2	0.8	19.57 kW h

4.1.2 Occupancy Scenarios

Two different occupancy scenarios are considered: adaptive and baseline. In the baseline scenario, a constant temperature scheduling between comfort and standby mode is applied. In the adaptive scenario, the eco mode is used. In the following, the different occupancy scenarios are considered in a case study.

Case Study

In this case study, a small office space with five rooms is evaluated (see Fig. 4.1). The five rooms have different room-individual space conditioning requirements, which are assumed in this case study (see Tab. 4.5).

Table 4.5: Case study, based on (Frahm et al. 2023)

Room	Function	Use case
1	Main office	Full-time office space, used during the working day from 8AM to 5PM, except during lunch break from 12AM to 1PM.
2	Second office	Part-time office space, used in the morning between 8AM and 12AM.
3	Storage, printer	Rarely used during the working day.
4	Kitchen	Mostly used during lunch from 12AM to 1PM.
5	Bathroom	Occasionally used during the working day.

To address the space condition requirements, three different control modes are available: comfort, eco, and standby (see Sec. 3.4). While the adaptive scenario applies all three available control modes depending on the occupancy schedule from the use case, the baseline scenario uses only the comfort and standby modes. In summary, the control modes with the temperature ranges from Tab. 3.1 on page 40 are the following:

- **Comfort mode**: OTS level I with $[22.6\,°\mathrm{C},\ 24.0\,°\mathrm{C}]$.
- **Eco mode**: OTS level III with $[20.7\,°\mathrm{C},\ 25.8\,°\mathrm{C}]$.
- **Standby mode**: OTS level off with $[16.0\,°\mathrm{C},\ 30.0\,°\mathrm{C}]$.

Baseline Scenario

The baseline scenario applies the comfort mode during working hours when the building is occupied (from 8AM to 5PM). When the building is unoccupied, the standby mode is used to save energy. The resulting schedule for the baseline scenario is shown in Tab. 4.6.

Table 4.6: baseline scenario control modes for different periods and rooms, obtained from (Frahm et al. 2023)

Period / room	1	2	3	4	5
8AM to 5PM	comfort	comfort	comfort	comfort	comfort
else	standby	standby	standby	standby	standby

Adaptive Scenario

In comparison to the baseline scenario, the adaptive scenario uses a more complex schedule, as presented in Tab. 4.7. The adaptive scenario also applies the eco mode during the working day to save additional energy. The eco mode is designed to reduce the OTS level in rooms that are less frequently occupied, e.g. bathroom or kitchen, as explained in Sec. 3.4 on page 40.

Table 4.7: Adaptive scenario control modes for different periods and rooms, obtained from (Frahm et al. 2023)

Period / room	1	2	3	4	5
8AM to 12AM	comfort	comfort	eco	eco	eco
12AM to 1PM	eco	eco	eco	comfort	eco
1PM to 5PM	comfort	eco	eco	eco	eco
else	standby	standby	standby	standby	standby

4.1.3 Data

The framework is being tested in Eggenstein-Leopoldshafen, Germany, during the heating season of 2022/2023. The evaluation employs weather and dynamic pricing data from nine weeks (Nov 28, 2022 - Feb 06, 2023), each reflecting varied heating demands and costs. A measurement frequency of $\Delta t = 15\,\text{min}$ is used. Weather data is sourced from the KIT EnergyLab's weather station, measuring the ambient air temperature T_a and solar radiation $\dot{q}_\text{s}$.

The dynamic electricity price for 76344 Eggenstein-Leopoldshafen is obtained from "aWATTar HOURLY". The aWATTar tariff is based on the time-varying German day-ahead market, "EPEX Spot DE", which is provided by their API (awa 2023). This final consumer price includes the day-ahead market price with additional charges of three percent, as well as constant 0.0858 €/kWh for grid usage, and 0.0563 €/kWh for levies, charges, and taxes (HOU 2023), as presented in

Eq. (4.1). Energy fed into the grid is not compensated in this tariff. As a result, the price for selling power is equal to zero: $\Omega_{\text{sell}} = 0$.

$$\Omega_{\text{buy}}[k] = \text{EPEX Spot DE } [k] \cdot 1.03 + 0.0858€ + 0.0563€ \tag{4.1}$$

Finally, two boxplots for the prices Ω_{buy} and ambient temperatures T_{a} are presented in Fig. 4.2 and 4.3, respectively. The solid orange lines represent the median values, whereas the dashed green lines show the mean values.

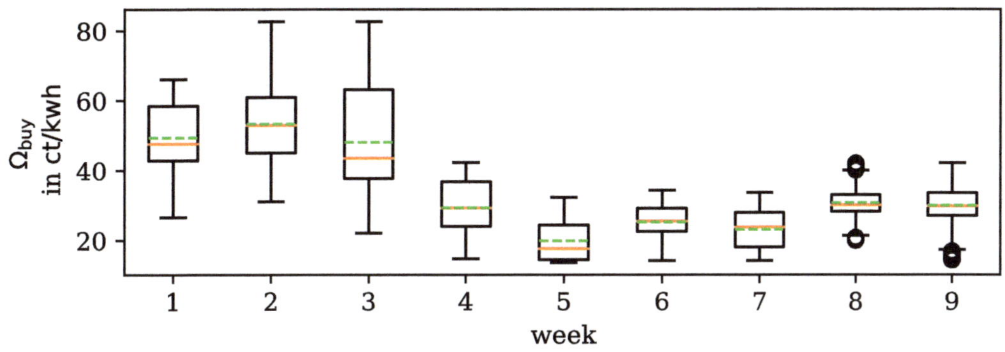

Figure 4.2: Overview about distributions of electricity prices across the different weeks

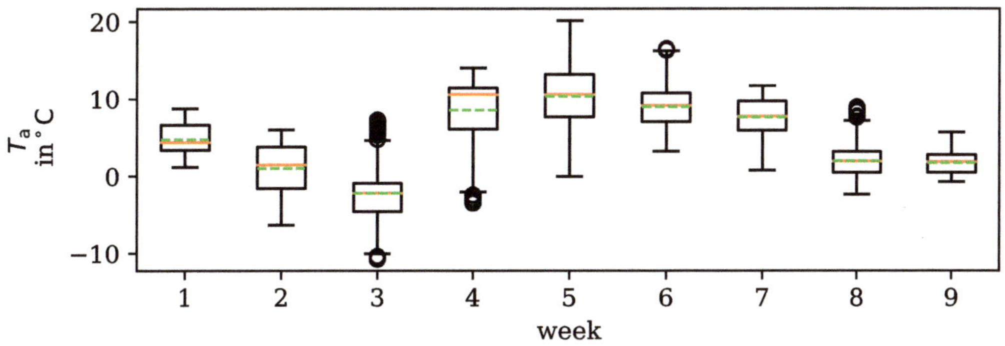

Figure 4.3: Overview about distributions of ambient temperatures across the different weeks

4.1.4 Uncertainties

The impact of uncertainties in the MPC planning process can be evaluated by comparing the results of Sec. 4.3.2 and 4.3.3. While the evaluations in both sections use the same ground truth, only the MPC in Sec. 4.3.3 is affected by uncertainties. Uncertainties in the MPC planning process occur due to weather forecasting errors and a mismatch between simulation and controller model ("model error"). Forecasting and model errors are simulated in Sec. 4.3.3 by adding noise to the MPC's forecast and model. This noisy MPC is evaluated on a simulation without noise, resulting in a difference to the optimal planning. For comparison, Sec. 4.3.2 presents the results of an optimal planning, where no noise is added to the MPC.

In all cases, MPC in this thesis is designed as a deterministic MPC (DMPC) that compensates uncertainties only over its repeated open-loop, resulting in a feedback-effect. DMPC delivers high performance in terms of costs and OTS, although other MPC approaches exist that explicitly incorporate uncertainties into their planning process, such as stochastic MPC (SMPC) or robust MPC (RMPC). However, RMPC and SMPC are designed to meet the OTS limits even under uncertainties by artificially shrinking the limits in the planning process, which also increases the costs (Langner et al. 2024b,a).

The noise used in Sec. 4.3.3 is simulated by Gauß distributions, as described by Langner et al. (2024b,a). Langner et al. (2024a) evaluated the plant-model mismatch and the forecast errors. The plant-model mismatch describes the noise of the indoor air temperature states T_{i_j} from one-step-ahead errors. The forecast errors results from the difference between weather forecasts data and actual measurements of the ambient temperature T_{a} and global solar radiation ϕ_{s}. Their uncertainty quantification conducted an analysis of kurtosis and skewness and showed that a Gauß distribution is a valid assumption to describe these differences, similarly to (Amadeh et al. 2022).

The uncertainty parameters and their standard deviations are summarized in Tab. 4.8. From the Gauß distributions, values are taken and added to the building model outputs $v_{T_{\mathrm{i}_j}}$ and to the weather forecast $w_{T_{\mathrm{a}}}$, $w_{\dot{q}_{\mathrm{s}}}$ during the MPC planning process. As a result of the weather dependency, also the heat pump's COP and the PV power generation are subject to uncertainties. All Gauß distributions use a mean value of zero, except the distribution describing the uncertainty in the solar radiation forecast. Based on the result from Langner et al. (2024a), the solar radiation forecast error has a mean value of -23.15 W m^{-2}. In addition, the solar uncertainty is only valid during the daytime, as during nighttime the sun has set.

Table 4.8: Uncertainty parameters, based on (Langner et al. 2024a)

Uncert. parameter	$w_{\dot{q}_s}$	w_{T_a}	$v_{T_{i_1}}$	$v_{T_{i_2}}$	$v_{T_{i_3}}$	$v_{T_{i_4}}$	$v_{T_{i_5}}$
Standard deviation	$69\,\mathrm{W\,m^{-2}}$	1.61 K	0.011 K	0.006 K	0.018 K	0.010 K	0.020 K

4.1.5 Metrics

The two optimization goals are (i) reducing energy costs while (ii) maintaining OTS. The metrics for costs and OTS violations are described by the units EUR per week and Kh per week. Both metrics should be as low as possible. Oldewurtel et al. (2012) defined the allowed OTS violation as 70 Kh per year, which is equivalent to 1.34 Kh per week. For clarity reasons, the unit "per week" is not shown in the result sections.

Costs The energy costs depend on the dynamic electricity price as well as the purchased and sold electric power. To calculate the costs, the formula is equivalent to the optimization goal in Eq. (3.8) on page 30, as recaptured in Eq. (4.2).

$$\text{costs}[k] = \dot{W}_{\text{buy}}[k] \cdot \Omega_{\text{buy}}[k] - \dot{W}_{\text{sell}}[k] \cdot \Omega_{\text{sell}}[k] \tag{4.2}$$

Discomfort For simplicity, the "OTS violations" are described by the term "discomfort" (discomf) in the following. The discomfort evaluates whether the actual room temperatures satisfied the allowed OTS ranges. These allowed temperature ranges are time-variant and room-individual (see Sec. 4.1.2). When an actual room temperature leaves its allowed range, the discomf value increases, as described in Eq. (4.3) – (4.5). To obtain the unit Kh, the sample rate, converted into h, needs to be multiplied.

$$\text{discomf}_j[k] = \Delta t \cdot \max\left\{0,\ \Delta T_{\text{lb}_j}[k],\ \Delta T_{\text{ub}_j}[k]\right\} \tag{4.3}$$

$$\Delta T_{\text{lb}_j}[k] = -T_{\text{i}_j}[k] + T_{\text{i,lb}_j}[k] \tag{4.4}$$

$$\Delta T_{\text{ub}_j}[k] = T_{\text{i}_j}[k] - T_{\text{i,ub}_j}[k] \tag{4.5}$$

Finally, the OTS violation is summarized over the five rooms j (see Eq. (4.6)).

$$\text{discomf}[k] = \sum_{j}^{n} \text{discomf}_j[k] \tag{4.6}$$

4.2 Comparison of the New Framework with other Control Strategies

To evaluate the performance of the new framework, the MPC-based control strategy is benchmarked against two other control strategies, which are based on rule-based control (RBC) and feedback control (FBC) (see Sec. 2.4.1). The comparison is performed in a configuration with only a heat pump, without PV or batteries, as only MPC can explicitly consider these elements. The implementations of RBC and FBC, used as benchmarks in this dissertation, are presented in the following.

The applied RBC strategy is called price storage control (PSC) and modulates the heat pump's power based on two factors: price and storage factor (Frahm et al. 2023). Both factors vary in the range of 0 to 1 and are multiplied to obtain the modulation degree (also in the range of 0 to 1). The price factor depends on the dynamic energy price and it is higher with lower prices. The storage factor calculates the reserve of the current state of inside air temperature, relative to its boundaries. The storage factor becomes higher when the temperature reserve reduces.

The FBC is a hysteresis-based two-point controller that heats when the lower temperature boundary is reached (Frahm et al. 2023). The heating is applied over a defined period and then, turned off until the boundary is reached again.

Overall, the three control strategies, MPC, RBC, and FBC, differ in their decision-making process and their controller complexity. MPC is an optimality- and model-based approach that considers temperature constraints and forecasts of time-variable parameters such as weather and price. PSC is a rule-based controller that applies rules based on the current price and storage, quantified by the difference between actual and permitted temperature. FBC is feedback-based and reacts only on the current temperature ranges.

4.2.1 Demonstration of Controller Behavior

To demonstrate the control behavior, the MPC-based framework is presented with two different forecast horizons (32h, 16h) and compared with RBC and FBC. The results are shown in Fig. 4.4 and 4.5 for two different weeks with the adaptive scenario and low thermal capacity. These two weeks were chosen because of their significantly different weather conditions and prices. Week 3 features cold weather and high prices, while week 5 showcases warm weather and low prices.

The following section describes the results of these figures. Firstly, the plot characteristics are explained, followed by an exploration of the disparities among the various weeks and control approaches, including MPC 32h, MPC 16h, RBC and FBC.

As for the plot description, Fig. 4.4 and 4.5 showcase time measured in days for each of the distinct weeks. The graph displays the air temperature T_{i_j} in each room j, the electricity consumption of the heat pump $\dot{W}_{\mathrm{h}}$, the weather conditions T_{a}, $\dot{q}_{\mathrm{s}}$, and the electricity price Ω_{buy}. The blue regions indicate the allowable temperature ranges for the air temperatures $[T_{\mathrm{lb}_j}(t), T_{\mathrm{ub}_j}(t)]$.

Weeks For the purpose of week-to-week comparison, the study focuses on weather conditions and electricity prices. In general, the electricity prices and the ambient temperatures vary significantly over the different weeks (compare Fig. 4.2 and 4.3), while the solar radiation shows relatively similar characteristics in both weeks. The radiation is zero during the night and reaches around 200 W/m^2 during the day. In contrast, the ambient temperature differs significantly in both weeks. The temperature is the lowest during week 3, ranging from -10 °C to 0 °C (mean: −2.2 °C), and the highest during week 5 (between 0 °C and 20 °C, mean: 10.3 °C). As a result, the heating of $\dot{W}_{\mathrm{h}}$ shows the highest values during week 3 (up to 5000 W and frequently activated) and the lowest during week 5 (up to 3000 W and often deactivated). The electricity price is the highest during week 3, often around 50-75 ct/kWh, and the lowest during week 9 (around 20 ct/kWh).

Control Strategies The four different control strategies MPC 32h, MPC 16h, RBC and FBC are compared. In general, the MPC 32h and 16h behave almost identically. Both versions heat when the electricity price is at low values, which leads to a "preheating" behavior. With MPC, the temperature is often close to the lower limit of the temperature restrictions. However, the preheating behavior is more pronounced with the larger prediction horizon. RBC also shows a tendency to preheat at lower electricity prices. However, the temperature levels for RBC are higher on average than for MPC. In contrast to MPC and RBC, FBC does not show an explicit response to the electricity price. Instead, the FBC only shows a switch-on and switch-off heating behavior depending on the temperature ranges.

4.2.2 Comparison of Control Performance

The metrics costs and OTS violations ("discomf") are used to evaluate control performance (see Sec. 4.1.5). The MPC-based framework is compared to RBC and FBC in two scenarios (baseline vs. adaptive) and on two different capacity types (high vs. low capacity) over nine weeks. In addition, the mean and standard deviation (SD) of costs and discomf are calculated over the nine

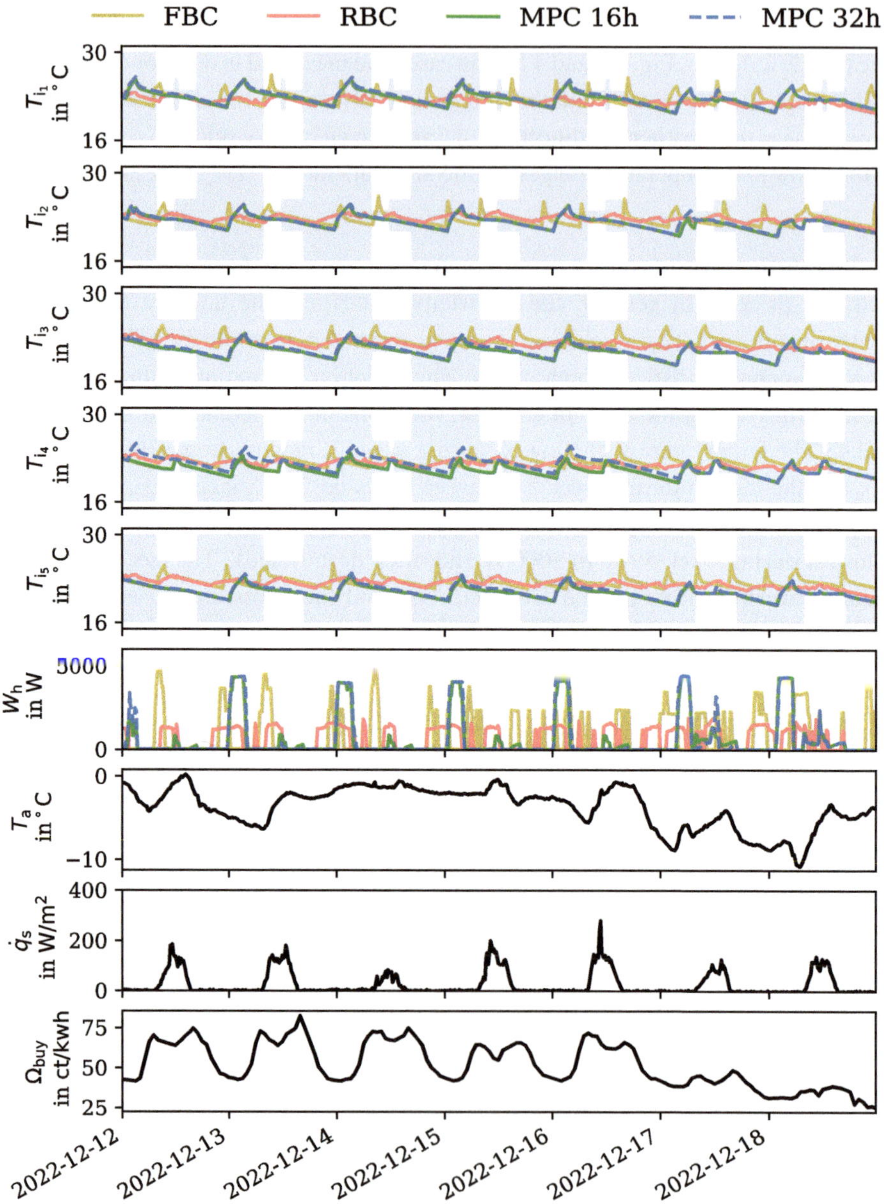

Figure 4.4: Control results in the adaptive scenario with low capacitance for week 3

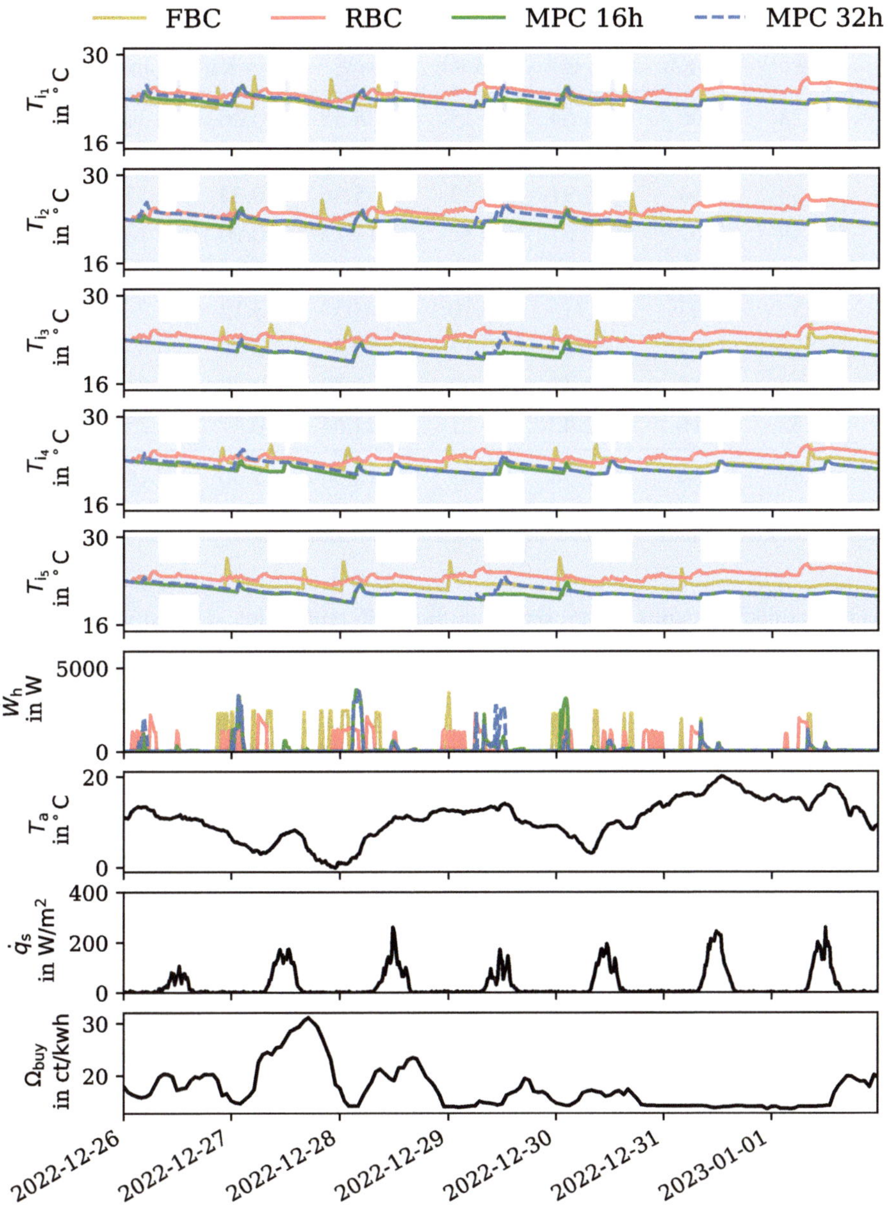

Figure 4.5: Control results in the adaptive scenario with low capacitance for week 5

Table 4.9: Comparison of control strategies with base scenario and low capacity

week	MPC 32h		MPC 16h		RBC		FBC	
	costs in €	discomf in Kh	costs in €	discomf in Kh	costs in €	discomf in Kh	costs in €	discomf in Kh
1	24.28	0.08	24.64	0.15	28.38	0.26	30.52	16.80
2	31.35	0.08	31.79	0.17	36.72	0.13	38.75	18.32
3	45.34	0.14	45.91	0.22	55.56	0.34	57.78	11.10
4	10.49	0.07	10.67	0.12	14.13	0.90	13.85	13.58
5	4.50	0.05	4.58	0.10	6.48	16.35	5.78	14.03
6	6.35	0.07	6.71	0.10	9.05	3.12	7.72	18.73
7	6.55	0.07	6.72	0.12	8.63	0.06	8.63	21.94
8	19.25	0.12	19.42	0.20	21.29	0.20	21.69	18.65
9	22.33	0.16	22.42	0.21	24.64	0.47	25.40	15.94
std	13.62	0.04	13.74	0.05	15.04	5.00	16.13	3.10
mean	18.94	0.09	19.21	0.15	22.76	2.43	23.35	16.57

weeks. The evaluation results are presented in four different tables, depending on scenario and capacity: Tab. 4.9 (base, low), Tab. 4.10 (adaptive, low), Tab. 4.11 (base, high), and Tab. 4.12 (adaptive high).

First, the results from Tab. 4.10 are evaluated, describing the adaptive scenario with low thermal capacity, similarly to Sec. 4.2.1. This evaluation describes the differences between weeks and controllers. Then, the different tables are compared to also present differences between capacities and scenarios.

Weeks The week. determining weather and energy prices, has a greater impact on costs than on discomf. Week 5 is associated with the lowest costs, whereas week 3 incurs the highest costs for all controllers, scenarios, and capacities. Tab. 4.10 outlines the basis for these observations. Similar trends are observed across varied capacities and scenarios. MPC 16h incurred costs of 3.76 € in week 5 and 42.00 € in week 3. Overall, the MPC 16h has mean costs of 17.18 € and a SD of 12.76 €. The discomf, on the other hand, remains relatively consistent throughout the different weeks. There is only one major exception for RBC in week 5 that shows relatively high discomfort (6.65Kh).

Table 4.10: Comparison of control strategies with adaptive scenario and low capacity

	MPC 32h		MPC 16h		RBC		FBC	
week	costs in €	discomf in Kh	costs in €	discomf in Kh	costs in €	discomf in Kh	costs in €	discomf in Kh
1	21.67	0.12	22.27	0.23	27.04	0.91	30.90	5.12
2	28.16	0.11	28.8	0.24	34.94	0.27	38.50	6.33
3	41.26	0.16	42.00	0.30	49.12	3.47	59.15	4.23
4	8.82	0.11	9.08	0.19	13.75	1.16	13.76	4.88
5	3.65	0.21	3.76	0.14	6.47	6.65	5.43	4.82
6	5.20	0.13	5.49	0.15	8.90	2.13	7.72	5.38
7	5.40	0.12	5.60	0.17	8.26	1.26	8.25	6.27
8	17.28	0.29	17.53	0.25	20.82	1.37	21.69	6.12
9	19.94	0.22	20.05	0.28	23.74	1.75	25.75	6.08
std	12.56	0.06	12.76	0.06	13.33	1.81	16.54	0.72
mean	16.82	0.16	17.18	0.22	21.45	2.11	23.46	5.47

Table 4.11: Comparison of control strategies with base scenario and high capacity

	MPC 32h		MPC 16h		RBC		FBC	
week	costs in €	discomf in Kh	costs in €	discomf in Kh	costs in €	discomf in Kh	costs in €	discomf in Kh
1	23.65	0.05	24.50	0.08	28.28	0.05	31.30	17.40
2	30.56	0.05	31.49	0.10	36.46	0.01	35.80	22.98
3	44.49	0.07	45.86	0.13	55.07	0.12	62.07	14.12
4	10.10	0.04	10.37	0.07	14.46	0.11	13.38	23.21
5	4.22	0.03	4.48	0.06	6.60	5.44	5.35	13.79
6	6.03	0.04	6.56	0.06	9.06	0.55	7.68	18.11
7	6.26	0.04	6.58	0.07	8.75	0.00	8.54	18.49
8	18.82	0.07	19.33	0.11	21.35	0.01	20.95	25.20
9	21.86	0.09	22.03	0.11	24.42	0.04	24.80	19.98
std	13.41	0.02	13.74	0.03	14.84	1.68	17.04	3.76
mean	18.44	0.05	19.02	0.09	22.72	0.70	23.32	19.25

Table 4.12: Comparison of control strategies with adaptive scenario and high capacity

	MPC 32h		MPC 16h		RBC		FBC	
week	costs in €	discomf in Kh	costs in €	discomf in Kh	costs in €	discomf in Kh	costs in €	discomf in Kh
1	20.44	0.10	21.32	0.15	26.99	0.88	31.09	7.90
2	26.35	0.09	27.85	0.16	34.84	0.02	35.84	7.29
3	38.91	0.11	40.45	0.18	46.92	3.19	64.97	4.42
4	7.99	0.10	8.26	0.12	13.92	0.15	12.85	7.58
5	3.06	0.16	3.21	0.09	6.55	2.74	5.34	4.37
6	4.60	0.08	4.94	0.09	9.10	0.51	8.82	5.32
7	4.84	0.09	5.00	0.11	8.50	0.35	8.72	5.90
8	16.11	0.19	16.74	0.17	20.88	0.46	20.68	8.30
9	18.45	0.14	19.13	0.19	23.18	0.98	26.05	5.84
std	11.95	0.04	12.46	0.04	12.74	1.08	17.70	1.40
mean	15.64	0.12	16.32	0.14	21.21	1.03	23.82	6.32

Control Strategies The MPC controller achieves superior performance over alternative control strategies in terms of costs and discomf. According to Tab. 4.10, the mean cost of MPC 32h is 16.82 €, which is 21.6 % less expensive than RBC (21.45 €) and 28.3 % cheaper than FBC (23.46 €). There are minor differences between MPC 16h and MPC 32h. The latter costs 2.1 % points less (16.82 € vs. 17.18 €). The discomf of both MPC versions is almost 0 Kh, which is approximately 2 Kh lower than RBC and 5 Kh lower than FBC. In view of the permissible OTS violation of 1.34 Kh/week, only MPC consistently meets this criterion.

Capacities While higher capacities lead to an improvement in cost using MPC, the impact on RBC and FBC is relatively low. When comparing Fig. 4.10 and 4.12, the costs of MPC 16h reduce by 5.0 % with the higher capacitance (from 17.18 € to 16.32 €). A cost reduction of 7.0 % is observed with MPC 32h (from 16.82 € to 15.64 €). With RBC, the costs decrease by 1.1 % (from 21.45 € to 21.21 €) and with FBC, the costs increase by 1.5 % (from 23.46 € to 23.82 €).

Scenarios Comparing Tab. 4.9 and 4.10 shows the differences between the baseline and adaptive scenario for low capacity. The adaptive scenario decreases the costs most significantly with MPC: 11.2 % for MPC 32h (16.82 € vs. 18.94 €), 10.6 % for MPC 16h (17.18 € vs. 19.21 €), 5.8 % for RBC (21.45 € vs. 22.76 €), and 0.5 % for FBC (23.46 € vs. 23.35 €).

4.2.3 Discussion

The results from Sec. 4.2.1 and 4.2.2 are discussed with focus on weeks, capacities, scenarios, and controllers.

Weeks The evaluation over multiple weeks enhances the significance of the results as single-week evaluations lack representation. Heating costs fluctuate significantly based on weather conditions and electricity prices across weeks. Using MPC, costs range from approximately 4 € to 40 € per week, with a mean and SD of approximately 18 € and 12 €, respectively.

Control Strategies MPC achieves optimal control performance with more than 20 % lower costs than conventional control strategies, while also improving OTS. These minimal costs and discomfort result from using an optimization algorithm, an underlying model, and the incorporation of time-variable parameters for its decision-making process. Not only can MPC integrate weather forecasts and electricity prices, but it can also predict the future system behavior to schedule heating loads optimally. This leads to achieving the lowest feasible indoor air temperature to save energy, while meeting the OTS. Additionally, the MPC exhibits pre-heating behavior during low energy prices, further reducing costs.

Capacities Thermal capacity indicates internal thermal energy storage. Greater capacitance provides more heat storage for the same temperature change. Temperature changes follow the proportionality $\Delta T \propto \frac{1}{C}\dot{Q}$, where C is capacitance and $\dot{Q}$ is heating. All controllers exploit this capacitance to adjust their heating over time while staying within permitted temperature ranges. However, the benefit of the capacity is the highest when using MPC. As MPC uses a model to determine the optimal heating trajectory it can explicitly exploit the thermal capacitance.

Scenarios The adaptive and baseline scenarios are designed to save energy during unattended periods. The scenarios differ in complexity when considering the time-variable occupancy. The adaptive scenario is more complex (see Sec. 4.1.2), which is intended to save the most energy. In the adaptive scenario, the MPC takes full advantage of the possibility to reduce the air temperature during unoccupied periods. This reduces the energy costs by 12 % compared to the baseline scenario, while still providing OTS. This shows the potential of room-individual thermal control and scheduling of time-variable constraints based on individual occupancy presence.

4.3 Evaluation of Different Framework Configurations

This section compares four framework configurations using a heat pump and incorporating PV and/or batteries (see Sec. 3.1.1). The four configurations are defined as follows:

(a) heat pump (equivalent to Sec. 4.2),

(b) heat pump and PV,

(c) heat pump and battery,

(d) heat pump, PV, and battery.

The evaluation is performed in the adaptive scenario, with low capacitance, and a prediction horizon of 16 h. The PV and battery parameters can be found in Sec. 4.1.1. As described in the latter section, the initial and end SoE of the battery is set to as 50 %.

Sec. 4.3.2 and 4.3.3 distinguish between certainty and uncertainty results. Sec. 4.3.2 neglects model inaccuracies and forecasting errors, while Sec. 4.3.3 adds noise to the MPC, resulting in a mismatch from the optimal solution (see Sec. 4.1.4). Since both simulations employ the same ground truth, the results can be compared. Tab. 4.13 and 4.14 present the results for certainty and uncertainty, respectively.

4.3.1 Demonstration of PV and battery

Fig. 4.6 and 4.7 demonstrate the use of PV and battery by illustrating the application of framework configuration **(d)** (heat pump ,PV, and battery). Fig. 4.6 and 4.7 present week 3 and 5, respectively, because of their significant differences in weather conditions and prices, similarly to Fig. 4.4 and 4.5 in Sec. 4.2.1. Week 3 represents cold temperatures and high heating demand, while week 5 is warm and therefore has lower heating energy demand.

The graph shows the air temperature in each room, the electricity consumption of the heat pump $\dot{W}_{\mathrm{h}}$, for both certainty and uncertainty. In 4.6 and 4.7, the weather conditions and prices are faded out as they are equivalent to Fig. 4.4 and 4.5. Instead, the PV power generation $\dot{W}_{\mathrm{PV}}$, grid interaction $\dot{W}_{\mathrm{grid}}$ and SoE are presented here.

In both weeks, there are only minor differences between certainty and uncertainty in terms of the trajectories of indoor air temperature, heating, grid interaction, and SoE. Although slight differences in the heat pump's power consumption occur, the impact on the indoor air temperature is insignificant. The differences in the temperature lead to no significant violations of the permitted temperature ranges.

4.3.2 Certainty Results

The results under neglecting of uncertainties are presented in Tab. 4.13. The results are evaluated with a focus on the different configurations and weeks.

Table 4.13: Certainty – Comparison of different framework configurations, neglecting uncertainties

	(a) HP		(b) HP & PV		(c) HP & battery		(d) HP. PV & batt.	
week	costs in €	discomf in Kh	costs in €	discomf in Kh	costs in €	discomf in Kh	costs in €	discomf in Kh
1	22.27	0.23	12.09	0.17	22.55	0.25	13.01	0.22
2	28.80	0.24	13.85	0.06	29.01	0.26	13.72	0.24
3	42.00	0.30	24.37	0.14	40.43	0.34	22.84	0.30
4	9.08	0.19	3.01	0.07	9.03	0.21	1.53	0.16
5	3.76	0.14	0.34	0.00	3.75	0.14	0.04	0.00
6	5.49	0.15	0.37	0.00	5.56	0.16	0.11	0.02
7	5.60	0.17	0.68	0.01	5.29	0.19	0.67	0.06
8	17.53	0.25	5.47	0.06	18.10	0.25	4.77	0.17
9	20.05	0.28	11.89	0.20	19.81	0.28	10.61	0.24
std	12.76	0.06	8.18	0.07	12.45	0.06	8.02	0.11
mean	17.18	0.22	8.01	0.08	17.06	0.23	7.48	0.16

Configurations Firstly, the mean results of the four different configurations are compared over all nine weeks (Tab. 4.13). While discomf is nearly zero for all four configurations, significant cost differences exist. The costs of the (a) heat pump configuration (17.18 €) can be decreased by 53.4 % with (b) PV (8.01 €), by 0.7 % with (c) battery (17.06 €), and by 56.5 % with (d) PV and battery (7.48 €).

Weeks For all the configurations, there is a significant variation in the costs over the nine weeks. The highest costs occur in week 3, while the lowest costs occur between weeks 5 and 7, resulting in significant standard deviations. The magnitude of the SD falls within the mean

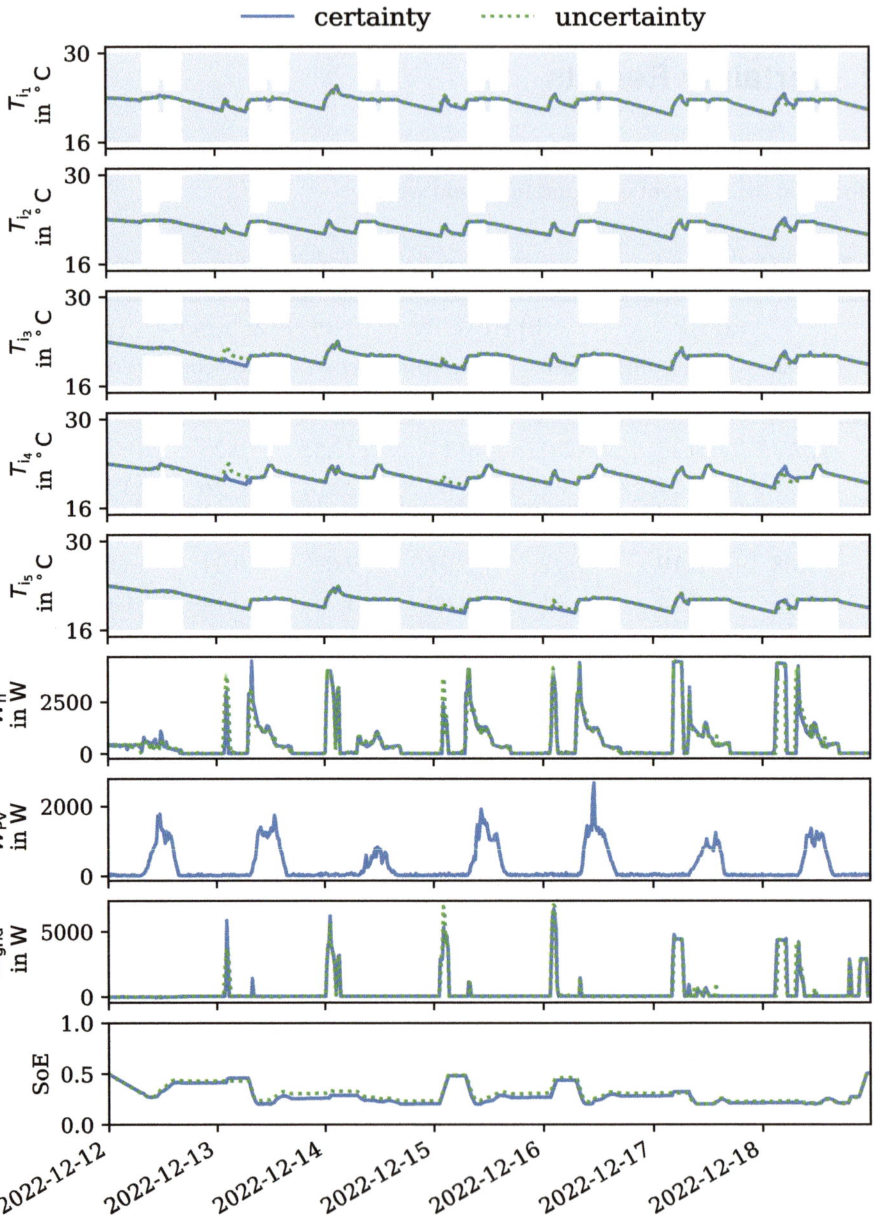

Figure 4.6: Component scheduling for configuration **(d)** in the adaptive scenario with low capacitance for week 3

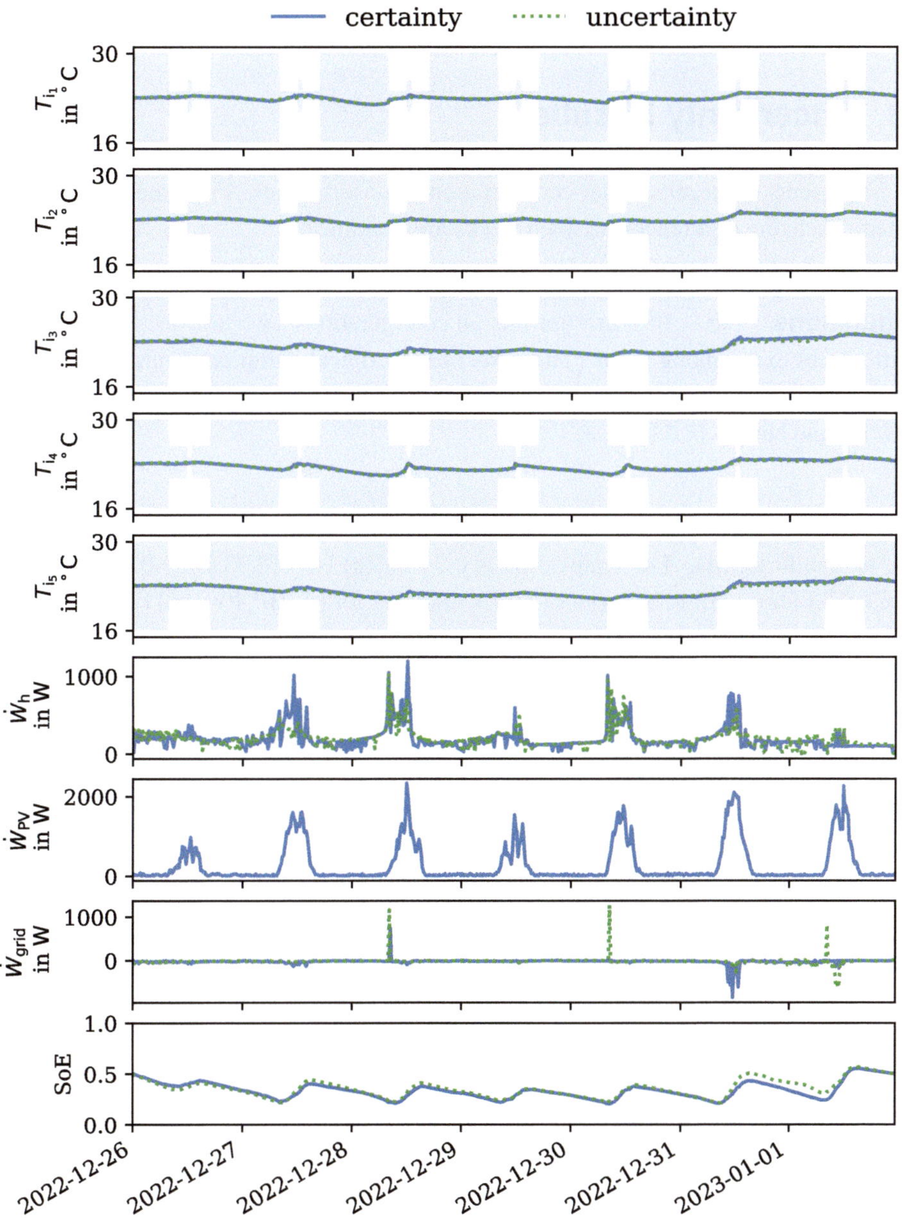

Figure 4.7: Component scheduling for configuration **(d)** in the adaptive scenario with low capacitance for week 5

value's range, leading to a relative standard deviation of 0.74 for (a) HP (12.76 €), 1.02 for (b) HP & PV (8.18 €), 0.73 for (c) HP & battery (12.45 €), and 1.07 for (d) HP, PV, and battery (8.02 €).

4.3.3 Uncertainty Results

Tab. 4.14 presents the results under consideration of uncertainties. First, the configurations are evaluated. Then, the differences between weeks are examined.

Configurations Tab. 4.14 illustrates that all configurations have low discomf, but significant cost differences exist among them. The costs of the (a) heat pump configuration (17.15 €) can be decreased by 54.5 % with (b) PV (7.81 €), by 0.3 % with (c) battery (17.09 €), and by 56.2 % with (d) PV and battery (7.51 €).

Weeks When comparing the different weeks, large differences in terms of costs can be observed, leading to significant SD. The relative SD is 0.74 for (a) HP (12.73 €), 1.02 for (b) HP & PV (7.97 €), 0.73 for (c) HP & battery (12.46 €), and 1.07 for (d) HP, PV, and battery (8.01 €).

Table 4.14. Uncertainty – Comparison of different framework configurations. considering uncertainties

	(a) HP		(b) HP & PV		(c) HP & battery		(d) HP. PV & batt.	
week	costs in €	discomf in Kh	costs in €	discomf in Kh	costs in €	discomf in Kh	costs in €	discomf in Kh
1	22.22	0.31	11.84	0.24	22.57	0.36	13.09	0.29
2	28.74	0.32	13.19	0.10	29.06	0.34	13.57	0.29
3	41.92	0.41	23.78	0.21	40.47	0.44	22.94	0.37
4	9.05	0.26	2.99	0.06	9.03	0.28	1.67	0.14
5	3.75	0.21	0.33	0.01	3.77	0.22	0.14	0.01
6	5.49	0.22	0.35	0.00	5.58	0.24	0.13	0.02
7	5.60	0.26	0.65	0.03	5.33	0.29	0.58	0.11
8	17.52	0.36	5.33	0.13	18.09	0.37	4.89	0.24
9	20.03	0.42	11.79	0.28	19.88	0.42	10.59	0.30
std	12.73	0.08	7.97	0.10	12.46	0.08	8.01	0.13
mean	17.15	0.31	7.81	0.12	17.09	0.33	7.51	0.20

4.3.4 Discussion

Weeks The evaluation over different weeks increases the significance of the following results, as they are averaged over various conditions. The weeks differ significantly in costs, which is a result of the different electricity prices and weather conditions. Lower temperatures and reduced solar radiation lead to increased heating demand, while solar radiation is the primary source of PV energy generation.

Configurations The discussion on configurations focuses on the impact of including PV or batteries in the MPC-controlled heat pump. When neglecting uncertainties, the inclusion of PV reduces operation costs by 53.4 %, the additional battery by 0.7 %, and the combination of PV and battery by 56.5 %. The cost savings of the PV is a result of the usage of solar energy, which minimizes the need for power purchase from the grid. The heat pump consumes this produced power and stores heat in the building. The battery can store additional energy, such as from PV generation or during times of low electricity prices, and discharge it during times of higher prices. The benefits of adding extra storage through batteries is limited for three reasons. The energy sold to the grid is not financially compensated for by the used aWATTar tariff. Secondly, considerable energy storage is already provided by the thermal energy storage of the building itself. Thirdly, the benefits of adding batteries are limited due to the losses during charging and discharging processes.

Uncertainties Due to uncertainties, the discomf slightly increases by up to 0.1 Kh/week, while the costs remain in the similar range. The increased discomf results from the model-plant-mismatch, as the MPC's internal model is subject to noise in the uncertainty case study. When the MPC controller aims to keep the temperature as close as possible to the lower OTS boundary (see Fig. 4.6 and 4.7), the noise may cause the temperature to fall below this boundary. However, the discomf remains under 0.5 Kh/week in all cases, which is below the permitted value of 1.34 Kh/week. Due to the time-varying OTS schedule, which relaxes temperature restrictions during unoccupied periods, the actual temperature often deviates from its temperature boundaries. Also, the plant-model mismatch has no significant impact on the costs, although minor differences in the heating trajectory occur due to noise. Similarly, the SoE trajectory from battery charging and discharging is slightly affected by errors in the weather forecast but this has no significant impact on the costs.

5 Experimental Model Evaluation

For the implementation of MPC in practical applications, a model is required. Previous publications pointed out the importance of multi-zone models in thermal building control and presented insights about different multi-zone structures (central vs. decentral) and different data qualities when using grey-box RC models (Frahm et al. 2022a), (Frahm et al. 2022c). However, conventional grey-box models are outperformed in accuracy by black-box SIM, which also have no difficulties with local optima and are computationally efficient (Yu et al. 2019, Verhaegen 2015). When using black-box identification, the models can also be divided into central and decentral approaches by explicitly allocating the data to the identification of different zones. Essentially, the black-box models are then designed similarly to the approaches described in Sec. 3.2 and applied in Sec. 4, except that no clear RC parameters are apparent. Instead, there are only system matrices.

This chapter presents a novel evaluation of multi-zone black-box SIM models identified from real data of an occupied residential building. SIM with different multi-zone structures are evaluated. In that context, a fusion multi-zone identification approach for residential buildings is proposed and investigated. Also, the importance of information quality of the data is evaluated by using nine weeks of identification and cross-validation over the remaining eight weeks in each case. Finally, this thesis proposes using the PARSIM-K algorithm in multi-zone building models as this algorithm is particularly suitable for closed-loop identification data, often present in actual building operations.

This chapter is organized as follows. Sec. 5.1 outlines the experimental setup, including the research infrastructure, model structures, algorithms, and data. Sec. 5.2 evaluates the results for identification and validation through objective error metrics. The results are summarized and discussed in Sec. 5.3.

5.1 Settings

The experimental setting is organized in software components and the infrastructure required for the field implementation. The software components consist of the model structures, algorithms, data, defined test cases, and defined error metrics. In addition to the software components developed in the scope of this thesis, a suitable infrastructure for a field implementation is presented in Sec. 5.1.1.

5.1.1 Infrastructure

As presented in Fig. 5.1, a real building provides the infrastructure to conduct the experiments. This building is part of the Karlsruhe Institute of Technology (KIT) EnergyLab (Hagenmeyer et al. 2016) and it is designed as a single-family home with five rooms, a kitchen, a bathroom, and a technical room. Each room is considered as a thermal zone with individual space-conditioning requirements. Currently, the building is utilized as office space and therefore occupied during working hours.

Figure 5.1: Research building for evaluation (left of the three houses)

The research building is equipped with a versatile energy system and several sensors[1]. The building generates heat with a heat pump and distributes it over radiators or underfloor heating. The power of the heat pump $\dot{W}_\mathrm{h}$ is measured as well as the temperatures in each thermal zone T_{i_j}. In addition, a weather station measures the ambient temperature T_a and global solar radiation $\dot{q}_\mathrm{s}$.

[1] Although the research building offers a versatile research environment, the installation is still an ongoing process. Therefore, the present thesis focuses on parts of the research infrastructure, that (a) suited the context of this work and (b) was sufficiently available during the time of processing this work.

5.1.2 Model Structures

As each room needs to be considered as its own zone, multi-zone thermal building model (MZTBM) structures are required (see Sec. 2.2.3). Three different MZTBM structures are compared: (i) central, (ii) decentral, and (iii) fusion MZTBM (see Sec. 3.2 and Frahm et al. (2022a,c)). The fusion MZTBM's parameters are identified from the average of the individual zone temperatures, like a single-zone model. However, the identified parameters of the fusion model are applied individually to each zone, similarly to the decentral approach. In contrast to the fusion model, the decentral model also identifies the parameters of each zone individually, using the individual air temperature in each zone. The decentral model and the fusion model consider no connections between adjacent zones. Only the central model considers connection between adjacent zones and identifies all parameters in a single connected structure.

5.1.3 Algorithms

The following three subspace identification methods (SIM) (see Sec. 2.4.5) are used: CVA, N4SID, and PARSIM-K. The major benefit of these SIM is that they need no parameterization of system matrices and no definition of start parameters. As a result, no problems due to nonlinear parameter optimization and local minima occur (Verhaegen 2015). PARSIM-K is explicitly designed for closed-loop identification in MPC design (Pannocchia and Calosi 2010a).

- **CVA**: Canonical Variate Analysis (Larimore 1990),
- **N4SID**: Numerical algorithms for Subspace State Space System Identification (Van Overschee and De Moor 1994),
- **PARSIM-K**: PARSIMonious subspace algorithm considering the system in predictor form (Pannocchia and Calosi 2010b).

All four identification algorithms are implemented in Python, using SIPPY (Armenise et al. 2018). The model order is limited by up to three states per zone.

5.1.4 Data

Nine weeks during the heating period 2022/2023 in Germany are used for identification and validation: Nov 28, 2022 - Feb 06, 2023. The identifications are always conducted over one entire week, from Monday to Sunday. During all weeks, the used data is considered as closed-loop data, where thermostats control the indoor air temperature, which can be manipulated by occupants.

Over most periods, the building is occupied. It remains only unoccupied on weekends and during winter holidays.

5.1.5 Test Cases

During each week, models are identified and cross-validated on the remaining eight weeks. For nine weeks of systems identification, this results in a total of 81 test simulations (9 identifications and 72 validations). These 81 simulations are performed on all three model types (decentral, central, and fusion) and three algorithms (N4SID, CVA, PARSIM-K), resulting in 729 simulations overall.

The results of the simulations are separated into identification and validation results. The identification results define the results for simulating the identified system with data of the same week, in which the system was identified (see Sec. 5.2.1). In contrast, validation results are the outcome of simulating on new data that was not involved during the identification process.

While the identification results describe how accurately the algorithm can find parameters under well-known conditions, the validation results indicate which identified system yields adequate performance over various, unknown, and changing system conditions. In the context of thermal building models, the change of conditions refers to weather and heating conditions as well as the occurrence of disturbances (e.g. open windows, open doors, shadings, internal heat gains).

For identification, one week is used, similarly to the literature (Arroyo et al. 2020, Drgoňa et al. 2021). For accurate day-ahead prediction in MPC, a validation period of 24 h is selected (Vallianos et al. 2022, Hauge Broholt et al. 2022), always at the beginning of the week. Beyont that, a 168 h ahead prediction is investigated for a one-week open-loop prediction. The latter evaluation is outside the focus scope of this thesis, and it is used to demonstrate the general prediction quality.

5.1.6 Metrics

Two different metrics are used to evaluate the performance of the identification and validation: mean absolute error (MAE) and root mean square error (RMSE) (Candanedo et al. 2018). Both metrics compare the simulation outputs $y_{i,j}[k]$ with the actual measurements $\hat{y}_{i,j}[k]$, as preseted in Eq. (5.1) and (5.2). The index i describes the week of evaluation, j the room, and k the step within the week.

$$\text{MAE} = \sum_{i}^{m}\sum_{j}^{n}\sum_{k}^{o} \frac{|y_{i,j}[k] - \hat{y}_{i,j}[k]|}{m \cdot n \cdot o} \tag{5.1}$$

$$\text{RMSE} = \sqrt{\sum_{i}^{m}\sum_{j}^{n}\sum_{k}^{o} \frac{(y_{i,j}[k] - \hat{y}_{i,j}[k])^2}{m \cdot n \cdot o}} \tag{5.2}$$

The measurements are considered as ground truth, although considerable disturbances occur due to real OB[2] and other measurement inaccuracies such as noise. For the MPC application, Hauge Broholt et al. (2022) define the required RMSE to be below 1 K over a predictive horizon of 24 h.

5.2 Evaluation Results

The identification and validation results are presented in Tab. 5.1 – 5.6. Tab. 5.1 and 5.2 show the identification and validation results for N4SID, Tab. 5.3 and 5.4 for CVA, and Tab. 5.5 and 5.6 for PARSIM-K.

The error metrics MAE and RMSE are used for all algorithms, structures, and weeks of identification. Additionally, the mean and SD are calculated for all nine identification weeks. The SD is used for the evaluation of outliers. Outliers describe results that differ significantly from the other results.

Sec. 5.2.1 presents the results during identification while Sec. 5.2.2 shows the results during 24-h validation. Sec. 5.2.3 outlines an additional 168-h validation. The results are initially described in a general manner and then analyzed further in terms of algorithms, structures, and data.

[2] These disturbances result from the use of windows, doors, shadings, appliances, and other internal gains.

Table 5.1: Identification results N4SID

	central		decentral		fusion	
ID week	MAE in K	RMSE in K	MAE in K	RMSE in K	MAE in K	RMSE in K
1	0.68	0.80	0.81	0.92	0.81	0.92
2	0.71	0.81	**2.66**	**2.90**	0.70	0.82
3	0.65	0.74	0.62	0.72	0.72	0.82
4	0.60	0.71	0.55	0.64	0.65	0.76
5	0.26	0.31	0.15	0.18	0.39	0.45
6	1.08	1.21	**1.37**	**1.62**	0.97	1.09
7	0.28	0.38	0.31	0.40	0.46	0.54
8	0.37	0.50	0.48	0.59	0.41	0.50
9	0.60	0.73	0.50	0.61	0.64	0.77
std	0.25	0.27	0.77	0.83	0.19	0.21
mean	0.55	0.65	0.82	0.94	0.59	0.69

Table 5.2: Validation results N4SID, 24h open-loop prediction

	central		decentral		fusion	
ID week	MAE in K	RMSE in K	MAE in K	RMSE in K	MAE in K	RMSE in K
1	0.56	0.64	0.52	0.60	0.49	0.56
2	0.74	0.85	**2.50**	**2.73**	0.47	0.59
3	0.67	0.72	0.44	0.51	0.39	0.46
4	0.60	0.71	0.58	0.72	0.56	0.70
5	0.70	0.85	0.66	0.81	0.64	0.81
6	0.57	0.65	**2.05**	**2.21**	0.51	0.58
7	0.51	0.61	0.58	0.68	0.54	0.65
8	0.54	0.61	0.71	0.84	0.46	0.52
9	1.15	1.38	1.08	1.32	1.10	1.35
std	0.20	0.24	0.75	0.79	0.21	0.27
mean	0.62	0.73	0.99	1.12	0.54	0.65

Table 5.3: Identification results CVA

	central		decentral		fusion	
ID week	MAE in K	RMSE in K	MAE in K	RMSE in K	MAE in K	RMSE in K
1	0.49	0.57	0.51	0.58	0.59	0.67
2	0.50	0.60	**2.75**	**3.07**	0.74	0.87
3	0.36	0.46	**2.27**	**2.72**	0.47	0.56
4	0.39	0.48	0.45	0.54	0.58	0.68
5	0.08	0.10	0.15	0.17	0.40	0.45
6	**3.49**	**4.28**	**1.34**	**1.59**	0.84	0.97
7	0.32	0.40	0.31	0.40	0.46	0.55
8	0.37	0.47	0.47	0.58	0.38	0.48
9	0.46	0.59	0.48	0.59	0.67	0.83
std	1.05	1.28	0.94	1.07	0.16	0.18
mean	0.75	0.92	0.97	1.13	0.53	0.62

Table 5.4: Validation results CVA, 24h open-loop prediction

	central		decentral		fusion	
ID week	MAE in K	RMSE in K	MAE in K	RMSE in K	MAE in K	RMSE in K
1	0.77	0.88	0.63	0.75	0.53	0.66
2	0.52	0.60	**2.70**	**2.94**	0.40	0.52
3	0.68	0.76	**5.06**	**5.33**	0.28	0.36
4	0.63	0.77	0.50	0.63	0.58	0.73
5	0.91	1.06	0.69	0.88	0.68	0.86
6	**8.60**	**9.62**	**2.11**	**2.29**	0.35	0.41
7	0.52	0.62	0.58	0.68	0.57	0.68
8	0.52	0.62	0.79	0.95	0.40	0.47
9	1.07	1.32	1.15	1.41	1.16	1.45
std	2.64	2.94	1.51	1.56	0.26	0.33
mean	1.69	1.92	1.57	1.74	0.52	0.65

Table 5.5: Identification results PARSIM-K

	central		decentral		fusion	
ID week	MAE in K	RMSE in K	MAE in K	RMSE in K	MAE in K	RMSE in K
1	0.52	0.60	0.48	0.57	0.47	0.58
2	0.56	0.69	0.49	0.57	0.49	0.61
3	0.58	0.66	0.61	0.69	0.66	0.74
4	0.31	0.41	0.29	0.38	0.46	0.52
5	0.27	0.31	0.14	0.16	0.39	0.44
6	0.52	0.61	0.48	0.55	0.52	0.59
7	0.28	0.39	0.29	0.40	0.46	0.55
8	0.36	0.50	0.36	0.47	0.41	0.52
9	0.58	0.71	0.62	0.74	0.68	0.82
std	0.13	0.14	0.16	0.18	0.10	0.12
mean	0.41	0.50	0.39	0.47	0.46	0.55

Table 5.6: Validation results PARSIM-K, 24h open-loop prediction

	central		decentral		fusion	
ID week	MAE in K	RMSE in K	MAE in K	RMSE in K	MAE in K	RMSE in K
1	0.53	0.60	0.42	0.50	0.41	0.47
2	0.66	0.75	0.51	0.65	0.50	0.63
3	0.68	0.74	0.58	0.65	0.57	0.64
4	0.46	0.54	0.33	0.40	0.33	0.40
5	0.71	0.86	0.63	0.81	0.63	0.81
6	0.65	0.72	0.66	0.74	0.68	0.76
7	0.49	0.59	0.39	0.48	0.36	0.45
8	0.55	0.62	0.61	0.72	0.50	0.57
9	0.71	0.86	0.72	0.90	0.67	0.84
std	0.10	0.12	0.13	0.16	0.13	0.16
mean	0.55	0.64	0.50	0.60	0.48	0.57

5.2.1 Identification Results

The identification results are presented in Tab. 5.1, 5.3, and 5.5 for the algorithms N4SID, CVA, and PARSIM-K, respectively. In all cases, mean MAE ranges from 0.39 K to 0.97 K, depending on structure and algorithm. Similarly, the mean RMSE ranges from 0.47 K to 1.13 K. Generally, the RMSE and MAE are correlated and differ in most cases by approximately 0.1 K. Because of this correlation, the following evaluation focuses only on one of these two metrics (using the MAE).

Algorithms The lowest identification errors can be achieved with the algorithm PARSIM-K, yielding a mean MAE between 0.39 K and 0.46 K, depending on the structure. The highest MAE is observed with CVA, ranging from 0.53 K to 0.97 K. With CVA, outliers can be found for the central and decentral structures, e.g. in week 6 with an MAE of 3.49 K and 1.34 K, respectively. N4SID shows a mean MAE between 0.55 K and 0.82 K and is also characterized by outliers. Outliers occur for N4SID with the decentral structure, e.g. in week 2 with a MAE of 2.66 K.

Structure The most consistent identification results can be obtained with the fusion structure, where the mean MAE ranges from 0.46 K to 0.59 K, depending on the algorithm. In contrast, the decentral structure shows the least consistent identification results, with a mean MAE from 0.39 K to 0.97 K. The decentral structure is subject to outliers when using the algorithms CVA or N4SID. The central structure has a mean MAE between 0.41 K and 0.75 K and shows outliers with the CVA algorithm.

Week The differences between weeks are evaluated for the algorithms and structures by using the standard deviation (SD) of the MAE. In terms of algorithms, PARSIM-K shows the lowest SD, ranging from 0.10 K to 0.16 K, depending on the structure. The highest SD occurs with CVA, where the SD ranges from 0.16 K to 1.05 K. N4SID has a SD between 0.19 K and 0.77 K. In terms of structures, the lowest SD can be found with the fusion structure, as the SD ranges from 0.1 K to 0.19 K. The remaining structures have similar SD: 0.16 K to 0.94 K for decentral and 0.13 K to 1.05 K for central.

Summary Overall, the algorithm PARSIM-K shows the lowest mean MAE, the lowest SD, and no outliers. The structure fusion yields the lowest SD and no outliers.

5.2.2 Validation Results

Tab. 5.2, 5.4, and 5.6 show the validation results for the algorithms N4SID, CVA, and PARSIM-K, respectively. Depending on algorithm and structure, the mean MAE ranges from 0.48 K to 1.69 K, while the mean RMSE ranges from 0.57 K to 1.92 K.

Algorithms The lowest validation errors can be achieved with the PARSIM-K algorithm, where the mean MAE ranges from 0.48 K to 0.55 K, depending on the structure. The highest errors occur with CVA, ranging from 0.52 K to 1.69 K. N4SID yields a mean MAE between 0.54 K and 0.99 K. The validation results of PARSIM-K are more consistent than with N4SID and CVA in terms of outliers. Outliers occur for N4SID with the decentral structure, e.g. in week 2 with a MAE of 2.50 K. Outliers can also be found for CVA with the central and decentral structures.

Structure The fusion structure yields the lowest errors and most consistent validation performance with a mean MAE ranging from 0.48 K to 0.54 K, depending on the algorithm. In contrast, consistency cannot be found for the central or decentral structures. For the central structure, the mean MAE is between 0.55 K to 1.69 K and outliers occur with the CVA algorithm. For the decentral structure, the mean MAE ranges from 0.50 K to 1.57 K and outliers can be found for N4SID and CVA.

Week The standard deviation (SD) of the MAE is used to characterize differences between weeks for the different algorithms and structures. The lowest SD can be found with the PARSIM-K algorithms, ranging from 0.10 K to 0.13 K, depending on the structure. In contrast, CVA shows the highest std of the MAE in the range of 0.26 K and 2.94 K. N4SID yields a SD between 0.54 K and 0.99 K. In terms of structures, the fusion structure yields the lowest SD, ranging from 0.13 K to 0.26 K, depending on the algorithm. The decentral structure has the highest SD with values between 0.13 K and 2.64 K. The SD of the central structure ranges from 0.10 K to 1.51 K.

Summary In summary, the validation results show similar findings to the identification results. The algorithm PARSIM-K yields the lowest mean MAE, the lowest SD, and no outliers. The structure fusion has the lowest SD and no outliers.

5.2.3 Larger Prediction Horizons

In this section, the algorithms and structures are evaluated over a 168-h open-loop predictive horizon to evaluate the general prediction quality. The results can be found in the appendix, in Tab. A.1, A.2, and A.3 for N4SID, CVA, and PARSIM-K, respectively.

Compared to the 24-h prediction, larger errors occur in the 168-h prediction. Nevertheless, several well-performing models can be found with validation errors of approximately 1 K MAE and RMSE (best case: 0.70 K MAE / 0.88 K RMSE). These well-performing models can be found for all types of algorithms and structures, but not for all weeks used for identification. In terms weeks used for identification, week 7 shows most adequate results for all cases with an MAE of approximately 1 K.

5.3 Discussion

The identification and validation performance depend on the structures, algorithms, and week used for identification. In general, the identification and 24h validation results show similar findings, where the algorithm PARSIM-K yields the lowest mean MAE, the lowest std, and no outliers. The structure fusion has the lowest SD and no outliers. Outliers occasionally occur in certain weeks for the remaining algorithms (N4SID, CVA) and structures (central, decentral). N4SID yields outliers with the decentral structure in week 2 and 6. CVA shows outliers with the decentral structure in week 2, 3, and 6 as well as for the central structure in week 6.

The most consistent predictive performance can be achieved with the PARSIM-K algorithm or the fusion structure. As PARSIM-K is explicitly designed for closed-loop identification, its consistent performance is reasonable. In case of the fusion structure, a consistent performance is a consequence of the model's high degree of generalization. This generalization results from averaging the indoor temperatures over different zones during identification. As a result, the impact of random disturbances in the individual zones, mostly related to occupant behavior, such as windows, shadings, or heat gains, is averaged over the zones.

The week used for identification has a significant impact on the validation results. The identification and 24h validation results showed that outliers occur only in certain weeks. The results from the 168h prediction show an even greater influence of the week used for identification. There are only a handful of weeks that lead to predictions with errors of about 1 K or below. For example, the identifications in week 7 yield the most adequate 168h predictions across all structures and algorithms.

Finally, from the knowledge gained, a recommendation is given on how a model for the MPC application should be selected. A combination needs to be chosen from a variety of algorithms, structures, and from the weeks of data that are available. Firstly, the selection of an algorithm and structure that are resistant to outliers can ensure adequate and consistent performance. Secondly, the chosen identification data should lead to a model that provides accurate prediction also for the validation during other weeks. In this study, the PARSIM-K algorithm and the fusion structure fulfil these requirements. When selecting a specific week of identification for PARSIM-K and fusion, week 7 demonstrates adequate validation performance with an MAE of 0.36 K and an RMSE of 0.45 K in the 24h prediction as well as 0.81 K and 0.97 K in the 168h prediction. The resulting model is exemplarily illustrated for three different 168h open-loop validation weeks in Fig. A.1, A.2, and A.3.

6 Conclusion

Building operations are responsible for 30 % of global final energy consumption (Internation Energy Agency 2023a), where about half of this energy use accounts for heating, ventilation, and air conditioning/cooling (HVAC) systems (Internation Energy Agency 2023b). The optimization of HVAC operations offers a large potential for energy optimization (Minoli et al. 2017), including the participation in demand response (DR) programs to handle the increasing share of renewable energies (Dengiz et al. 2019). Nevertheless, this energy consumption is directly linked to occupants, seeking comfortable conditions (Hong et al. 2017, Janda 2011). As a result, building control systems need to consider occupants' thermal satisfaction (OTS) room individually when optimizing energy use.

Chapter 1 presents the motivation for the thesis and the related work in the field of occupant-oriented demand response with multi-zone thermal building control and distributed energy sources. From the related work, research gaps are identified. The research gaps include the evaluation of different occupancy considerations and the integration of distributed energy sources such as photovoltaics (PV) and battery systems in multi-zone building control. Therefore, this thesis answers the research question on how to provide DR in a multi-zone residential building with distributed energy sources while considering room-individual occupancy. To answer the research question, a novel framework for occupant-oriented demand response is developed.

Chapter 2 establishes the fundamentals for the development of the novel framework, starting with laws of heat transfer and thermodynamics as fundamentals for model development and OTS consideration. Different modeling approaches are presented for multi-zone thermal building models, such as data-driven and physical or central and decentral. The basics of OTS are derived based on occupancy and parameters for thermal satisfaction. Finally, building control algorithms are introduced, including conventional control strategies, model predictive control (MPC), and the Kalman filter.

Chapter 3 introduces the novel framework based on MPC that optimizes the heat pump operation while accounting for PV, batteries, DR, and a time-variable schedule for OTS. The framework consists of six major parts: building model, occupant behavior (OB), constraints, predictions, cost function, and optimization. Significant attention is given to multi-zone building control and identification of multi-zone systems. Therefore, four different model structures are explained in detail: decentral, central, single-zone, and a novel fusion approach. Finally, an OTS model is developed that can be used for time-variable and room-individual occupancy consideration.

Chapter 4 evaluates the novel MPC-based framework in detail. It compares the developed framework to conventional control strategies and evaluates it in different configurations and under uncertainties in the prediction. The framework is evaluated in two different occupancy scenarios: baseline and adaptive, which vary in complexity. Additionally, the framework is examined in four different configurations, with different combinations of PV and battery elements. Finally, it is investigated how the framework can handle uncertainties by adding noise to the MPC's model and forecasts.

Chapter 5 investigates data-driven multi-zone thermal building models in a real, occupied building as a foundation for a field implementation of the framework. Three distinct multi-zone model structures are identified with three different black-box state space identification algorithms. These structures include a central and decentral approach, as well as a novel fusion structure. The black-box state space identification algorithms utilized in this study include N4SID, CVA, and PARSIM-K, with the latter being specifically suitable for closed-loop identification data. For identification and validation processes, nine weeks of winter data are used. Each identified model is cross-validated for the remaining eight weeks.

Most Important Results and Findings

1. Development of a novel framework that optimizes the heat pump operation while accounting for PV, batteries, DR, and time-variable schedule for OTS. Optimization of energy costs under consideration of time-variable constraints.
2. Evaluation of general control performance by comparison to conventional control strategies. The novel MPC-based framework reduces operational costs by more than 20 %, while also improving OTS.
3. Comparison of operational costs of the proposed control strategy for adaptive and baseline occupancy in multi-zone thermal building control. Adaptive occupancy consideration reduces costs by more than 10 %, compared to the baseline schedule.
4. Comparison of different framework configurations to evaluate the additional benefit of PV and batteries when already participating in a DR program. Integration of PV yields more than 50 % lower operation costs. Batteries yield only minor additional cost savings as thermal storage is already exploited by DR.

5. Uncertainty analysis in multi-zone control with DR and distributed energy sources by adding noise to the model and forecasts. Under uncertainties, the discomfort increases insignificantly (by about 0.1 Kh), while the costs remain similar.
6. Identification of multi-zone thermal building models in a real, occupied building. Identification with nine weeks and cross-validation over the remaining eight weeks in each case. Weeks used for identification have the most significant impact on validation results (compared to algorithms and structures).
7. Novel comparison of black-box state-space models with different multi-zone structures. Development of a novel fusion multi-zone identification approach for residential buildings. Fusion structure yields the most consistent validation results and lowest validation errors.
8. Novel evaluation of PARSIM-K algorithm in multi-zone building models that is suitable for closed-loop identification data as present in actual building operations. PARSIM-K yields the most consistent validation results in combination with the lowest validation errors.
9. Recommendation on how to select a model for the field implementation. Identification in week 7 with PARSIM-K and fusion yield in the validation an MAE of 0.36 K and RMSE of 0.45 K in the short term prediction (24h) as well as a 0.81 K MAE and 0.97 K RMSE in the long term prediction (168h).

Summary and Outlook

In summary, the results demonstrate a versatile applicability of the framework and its high performance in terms of operational costs and OTS. The modeling results indicate a simple yet accurate approach using the fusion structure or PARSIM-K algorithm, which is also applicable to closed-loop identification data. The control results show a high performance compared to conventional strategies and a high modularity, such as the integration of renewable energies, occupant orientation, and DR participation.

In future work, models can automatically be identified and adapted from measured data. The framework can be scaled from the individual building to the district level, allowing centralized distinct energy consumption management. Future building system configurations can also be taken into account, e.g. electric vehicles or future alternative heating technologies.

A Appendix

A.1 168 hours prediction

Table A.1: Validation results N4SID, 168h open-loop prediction

	central		decentral		fusion	
ID week	MAE in K	RMSE in K	MAE in K	RMSE in K	MAE in K	RMSE in K
1	1.33	1.55	2.44	2.85	2.02	2.34
2	2.67	3.16	4.25	4.66	2.12	2.49
3	**1.05**	**1.20**	1.84	2.17	1.64	1.93
4	2.49	2.90	2.74	3.17	2.68	3.11
5	2.60	3.02	2.78	3.22	2.75	3.18
6	2.13	2.49	3.74	4.11	2.61	3.03
7	**0.94**	**1.10**	1.25	1.48	**1.04**	**1.21**
8	**0.93**	**1.10**	3.70	4.29	**0.70**	**0.89**
9	7.00	8.00	6.36	7.23	6.86	7.81
std	1.89	2.16	1.51	1.67	1.79	2.02
mean	2.30	2.67	3.06	3.49	2.42	2.80

Table A.2: Validation results CVA, 168h open-loop prediction

	central		decentral		fusion	
ID week	MAE in K	RMSE in K	MAE in K	RMSE in K	MAE in K	RMSE in K
1	2.09	2.35	3.24	3.71	2.89	3.32
2	**0.88**	**1.07**	4.38	4.86	2.30	2.76
3	2.02	2.47	6.92	7.51	**0.91**	**1.11**
4	2.62	3.02	2.11	2.47	2.55	2.96
5	2.59	2.97	3.02	3.49	3.03	3.50
6	11.74	12.57	4.00	4.42	1.83	2.17
7	1.11	1.30	1.25	1.47	1.22	1.41
8	**0.81**	**1.03**	4.92	5.68	**0.70**	**0.88**
9	3.80	4.25	5.95	6.71	7.03	7.98
std	3.39	3.58	1.80	1.95	1.90	2.13
mean	3.10	3.46	3.76	4.23	2.44	2.82

Table A.3: Validation results PARSIM-K, 168h open-loop prediction

	central		decentral		fusion	
ID week	MAE in K	RMSE in K	MAE in K	RMSE in K	MAE in K	RMSE in K
1	**1.01**	**1.19**	1.47	1.76	**0.92**	**1.14**
2	1.16	1.37	1.83	2.17	1.23	1.47
3	2.91	3.30	3.20	3.67	3.21	3.67
4	**0.79**	**0.95**	**0.78**	**0.95**	**0.82**	**1.00**
5	2.74	3.16	2.75	3.18	2.69	3.11
6	3.74	4.31	4.40	5.11	4.50	5.22
7	**0.94**	**1.11**	**0.87**	**1.04**	**0.81**	**0.97**
8	1.51	1.76	3.12	3.62	2.22	2.58
9	3.67	4.21	3.89	4.50	3.90	4.51
std	1.21	1.37	1.30	1.49	1.41	1.60
mean	1.97	2.27	2.36	2.75	2.17	2.53

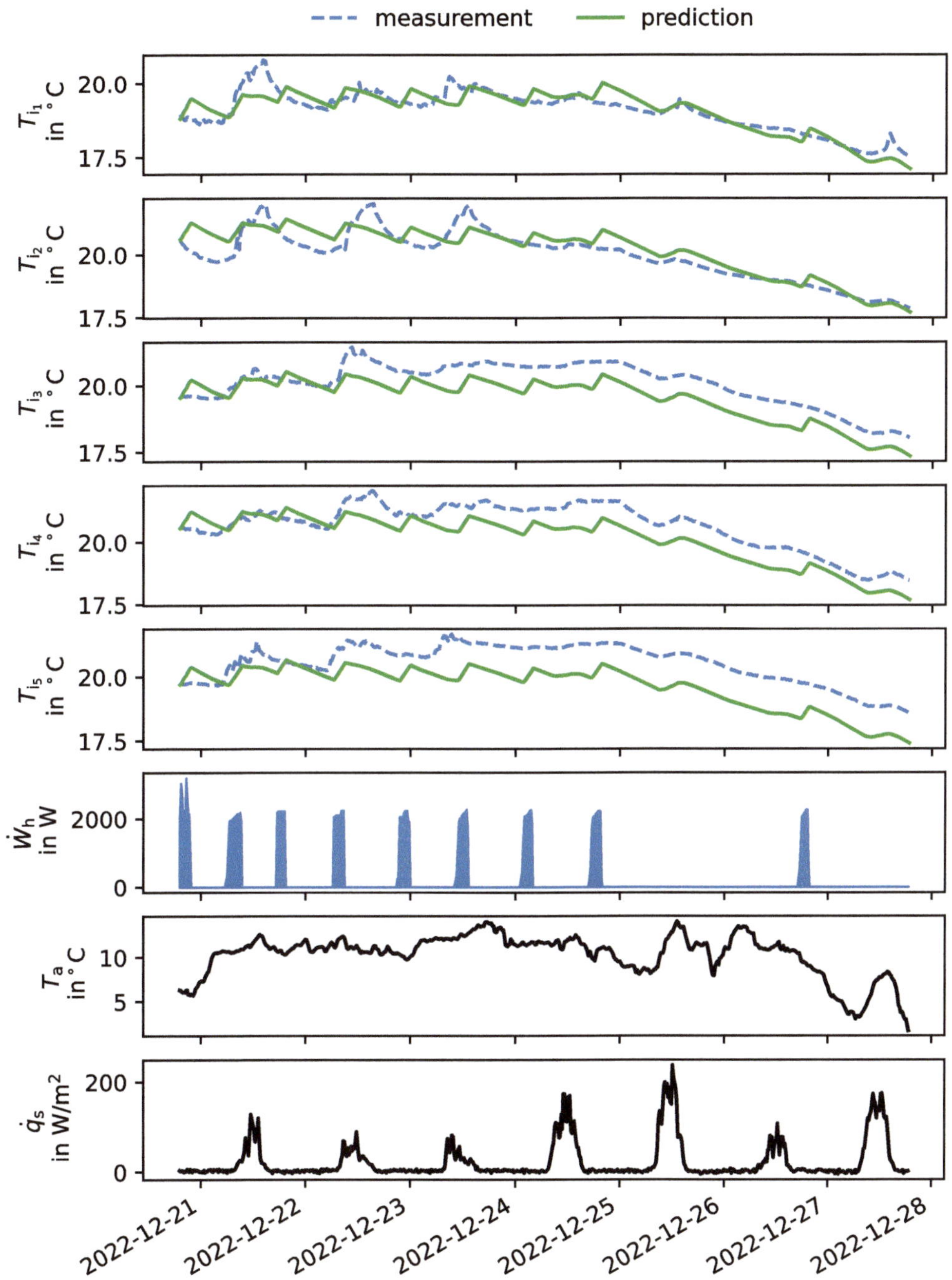

Figure A.1: Validation results of the fusion model, identified in week 7 and validated on week 4

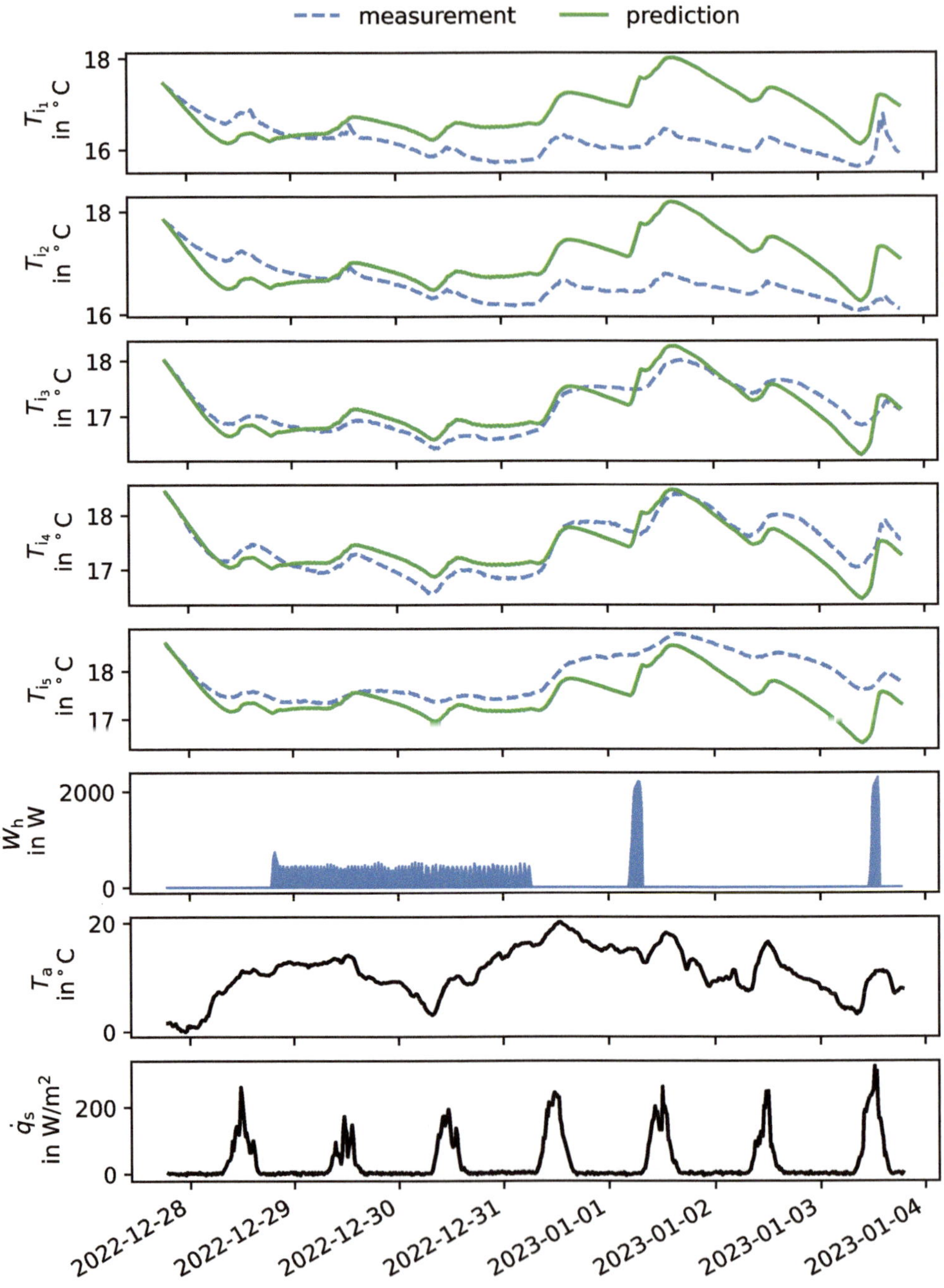

Figure A.2: Validation results of the fusion model, identified in week 7 and validated on week 5

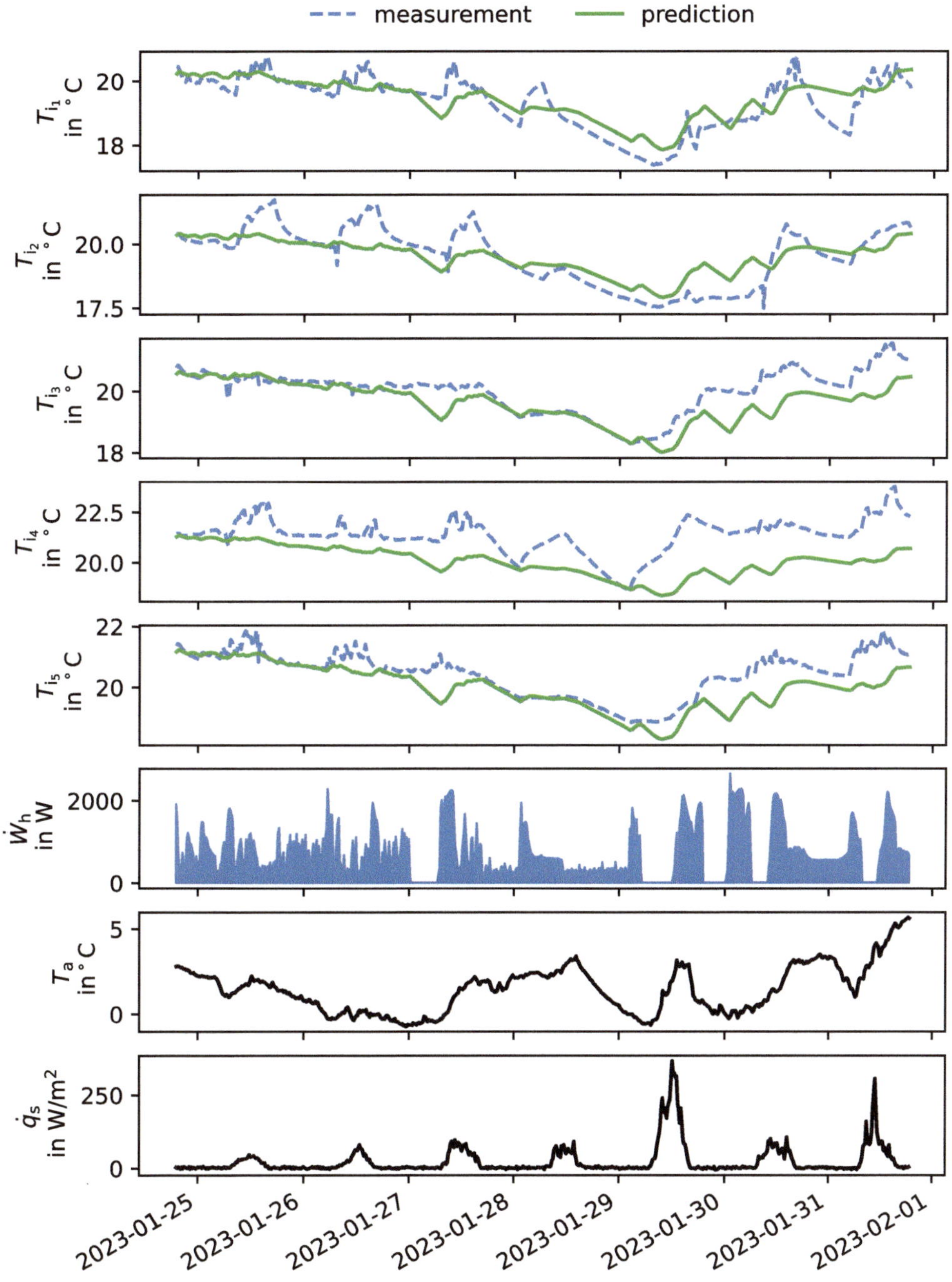

Figure A.3: Validation results of the fusion model, identified in week 7 and validated on week 9

List of Abbreviations

BEMS	building energy management system
COP	coefficient of performance
DMPC	deterministic MPC
DR	demand response
FBC	feedback control
HVAC	heating, ventilation, and air conditioning/cooling
KIT	Karlsruhe Institute of Technology
LTI	linear time-invariant
MAE	mean absolute error
MIMO	multiple input multiple output
MPC	model predictive control
MZTBM	multi-zone thermal building model
OB	occupant behavior
OTS	occupants' thermal satisfaction
PMV	predicted mean vote
PPD	predicted percentage of dissatisfied
PSC	price storage control
PV	photovoltaics
RBC	rule-based control
RC	resistor–capacitor

RMPC robust MPC

RMSE root mean square error

SD standard deviation

SIM subspace identification methods

SMPC stochastic MPC

SoE state of energy

List of Symbols

Indices

0	initial
a	ambient
batt	battery
buy	buying
cel	PV cell
ch	charging
dc	discharging
h	heat pump
I&C	inverter and controller
j	room index
kin	kinetic
lb	lower bound
max	maximal
min	minimal
module	PV module
noct	nominal operating cell temperature
op	operative
pot	potential
PV	photovoltaics
s	solar
sell	selling
stc	standard testing condition

T	temperature-dependent
ub	upper bound

Parameters

χ_{mod}	modulation degree
$\dot{q}$	radiation in $\mathrm{W\,m^{-2}}$
η	efficiency
Ω	dynamic electricity price in €/kWh
ε	coefficient of performance
a	surface in $\mathrm{m^2}$
C_{i}	heat capacity of room air in $\mathrm{J\,K^{-1}}$
C_{m}	heat capacity of heat accumulating medium in $\mathrm{J\,K^{-1}}$
g	gravitation in $\mathrm{N\,kg^{-1}}$
g_{s}	solar heat gain factor in $\mathrm{m^2}$
k	time step
N	number of steps
n	number of rooms j
p_{tv}	time-variable parameter
R_{a}	resistance between T_{i_j} and T_{a} in $\mathrm{K\,W^{-1}}$
R_{i}	resistance between T_{i} and T_{m} in $\mathrm{K\,W^{-1}}$

Variables

$\dot{m}$	mass flow in $\mathrm{kg\,s^{-1}}$
$\dot{Q}$	heat flow in W
$\dot{W}$	work flow in W
c	speed in $\mathrm{m\,s^{-1}}$
E	energy in J
h	specific enthalpy in $\mathrm{J\,kg^{-1}}$
S	entropy in $\mathrm{J\,K^{-1}}$

T	temperature in °C
t	time in s
T_{i}	room air temperature in °C
T_{m}	heat accumulating medium temperature in °C
U	internal energy in J
u	control input
v	output noise
w	process noise
x	state
y	control output
z	height in m

List of Figures

List of Tables

Bibliography

TARIF HOURLY - Deutschlands erster Stromtarif mit stündlicher Preisanpassung, 2023. URL `https://www.awattar.de/tariffs/hourly`. (accessed: 24.11.2023).

API - PREIS DATENFEED, 2023. URL `https://www.awattar.de/services/api`. (accessed: 24.11.2023).

Mohammed Jasim M. Al Essa. Home energy management of thermostatically controlled loads and photovoltaic-battery systems. *Energy*, 176:742–752, 2019. doi: https://doi.org/10.1016/j.energy.2019.04.041.

M. H. Albadi and E. F. El-Saadany. Demand Response in Electricity Markets: An Overview. In *2007 IEEE Power Engineering Society General Meeting*, pages 1–5, Tampa, FL, USA, June 2007. IEEE. doi: 10.1109/PES.2007.385728.

Ali Amadeh, Zachary E. Lee, and K. Max Zhang. Quantifying demand flexibility of building energy systems under uncertainty. *Energy*, 246:123291, May 2022. doi: 10.1016/j.energy.2022.123291.

ANSI/ASHRAE. Standard 55, Thermal Environmental Conditions for Human Occupancy, 2017.

Giuseppe Armenise, Marco Vaccari, Riccardo Bacci Di Capaci, and Gabriele Pannocchia. An Open-Source System Identification Package for Multivariable Processes. In *2018 UKACC 12th International Conference on Control (CONTROL)*, pages 152–157, September 2018. doi: 10.1109/CONTROL.2018.8516791.

Javier Arroyo, Fred Spiessens, and Lieve Helsen. Identification of multi-zone grey-box building models for use in model predictive control. *Journal of Building Performance Simulation*, 13 (4):472–486, July 2020. ISSN 1940-1493. doi: 10.1080/19401493.2020.1770861.

Ercan Atam and Lieve Helsen. Control-Oriented Thermal Modeling of Multizone Buildings: Methods and Issues: Intelligent Control of a Building System. *IEEE Control Systems Magazine*, 36(3):86–111, June 2016. ISSN 1941-000X. doi: 10.1109/MCS.2016.2535913. Conference Name: IEEE Control Systems Magazine.

Hans Dieter Baehr and Stephan Kabelac. *Thermodynamik (16. Edition)*. Springer Vieweg Berlin, Heidelberg, 2016. doi: https://doi.org/10.1007/978-3-662-49568-1.

Emrah Biyik and Aysegul Kahraman. A predictive control strategy for optimal management of peak load, thermal comfort, energy storage and renewables in multi-zone buildings. *Journal of Building Engineering*, 25:100826, September 2019. doi: 10.1016/j.jobe.2019.100826.

David H. Blum, Nora Xu, and Leslie K. Norford. A novel multi-market optimization problem for commercial heating, ventilation, and air-conditioning systems providing ancillary services using multi-zone inverse comprehensive room transfer functions. *Science and Technology for the Built Environment*, 22(6):783–797, August 2016. doi: 10.1080/23744731.2016.1197718.

Stephen P. Boyd and Lieven Vandenberghe. *Convex optimization*. Cambridge University Press, Cambridge New York Melbourne New Delhi Singapore, version 29 edition, 2023.

Luis M. Candanedo, Véronique Feldheim, and Dominique Deramaix. Reconstruction of the indoor temperature dataset of a house using data driven models for performance evaluation. *Building and Environment*, 138:250–261, June 2018. doi: 10.1016/j.buildenv.2018.04.035.

Salvatore Carlucci, Marilena De Simone, Steven K. Firth, Mikkel B. Kjærgaard, Romana Markovic, Mohammad Saiedur Rahaman, Masab Khalid Annaqeeb, Silvia Biandrate, Anooshmita Das, Jakub Wladyslaw Dziedzic, Gianmarco Fajilla, Matteo Favero, Martina Ferrando, Jakob Hahn, Mengjie Han, Yuzhen Peng, Flora Salim, Arno Schlüter, and Christoph Van Treeck. Modeling occupant behavior in buildings. *Building and Environment*, 174:106768, May 2020. doi: 10.1016/j.buildenv.2020.106768.

CEN. EN 16798-1 - Energy performance of buildings - Ventilation for buildings. Part 1: Indoor environmental input parameters for design and assessment of energy performance of buildings addressing indoor air quality, thermal environment, lighting and acoustics, 2019.

Jiří Cígler, Dimitrios Gyalistras, Jan Široky, V Tiet, and Lukaš Ferkl. Beyond theory: the challenge of implementing model predictive control in buildings. In *Proceedings of 11th Rehva world congress, Clima*, volume 250, 2013.

J A Crabb, N Murdoch, and J M Penman. A simplified thermal response model. *Building Services Engineering Research and Technology*, 8(1):13–19, February 1987. doi: 10.1177/014362448700800104.

Roel De Coninck and Lieve Helsen. Practical implementation and evaluation of model predictive control for an office building in Brussels. *Energy and Buildings*, 111:290–298, January 2016. doi: 10.1016/j.enbuild.2015.11.014.

Thomas Dengiz, Patrick Jochem, and Wolf Fichtner. Demand response with heuristic control strategies for modulating heat pumps. *Applied Energy*, 238:1346–1360, March 2019. doi: 10.1016/j.apenergy.2018.12.008.

Ján Drgoňa, Javier Arroyo, Iago Cupeiro Figueroa, David Blum, Krzysztof Arendt, Donghun Kim, Enric Perarnau Ollé, Juraj Oravec, Michael Wetter, Draguna L. Vrabie, and Lieve Helsen. All you need to know about model predictive control for buildings. *Annual Reviews in Control*, 50:190–232, 2020. doi: 10.1016/j.arcontrol.2020.09.001.

Ján Drgoňa, Aaron R. Tuor, Vikas Chandan, and Draguna L. Vrabie. Physics-constrained deep learning of multi-zone building thermal dynamics. *Energy and Buildings*, 243:110992, July 2021. doi: 10.1016/j.enbuild.2021.110992.

Norbert Elsner and Achim Dittmann. *Band 1 Energielehre und Stoffverhalten*. De Gruyter, Berlin, Boston, 1993. doi: doi:10.1515/9783112754979. URL `https://doi.org/10.1515/9783112754979`.

P. O. Fanger. Thermal comfort. Analysis and applications in environmental engineering. *Thermal comfort. Analysis and applications in environmental engineering.*, 1970.

Lukáš Ferkl and Jan Široký. Ceiling radiant cooling: Comparison of ARMAX and subspace identification modelling methods. *Building and Environment*, 45(1):205–212, January 2010. doi: 10.1016/j.buildenv.2009.06.004.

Christian Finck, Rongling Li, and Wim Zeiler. Economic model predictive control for demand flexibility of a residential building. *Energy*, 176:365–379, June 2019. doi: 10.1016/j.energy.2019.03.171.

Moritz Frahm. Model predictive control in buildings: Accounting for uncertainties of weather and occupancy behavior forecasts on demand-side-management. In *Annex 79 ST2 PhD Forum*, 2021. doi: 10.5445/IR/1000140832.

Svenne Freund and Gerhard Schmitz. Implementation of model predictive control in a large-sized, low-energy office building. *Building and Environment*, 197:107830, June 2021. doi: 10.1016/j.buildenv.2021.107830.

Freund, S. and Schmitz, G. Entwicklung und Validierung von Grey-Box-Modellen zur Modellierung des thermischen Verhaltens von Einzelbüros in einem Niedrigenergie-Bürogebäude. In *BauSIM*, 2020. doi: 10.3217/978-3-85125-786-1-55.

Andreas Geiger, Joachim Benner, Karl-Heinz Häfele, and Veit Hagenmeyer. Thermal Energy Simulation of Buildings based on the CityGML Energy Application Domain Extension. *BauSIM*, page 295, 2018.

Veit Hagenmeyer, Hüseyin Kemal Çakmak, Clemens Düpmeier, Timm Faulwasser, Jörg Isele, Hubert B. Keller, Peter Kohlhepp, Uwe Kühnapfel, Uwe Stucky, Simon Waczowicz, and Ralf Mikut. Information and Communication Technology in Energy Lab 2.0: Smart Energies System Simulation and Control Center with an Open-Street-Map-Based Power Flow Simulation Example. *Energy Technology*, 4(1):145–162, 2016. doi: 10.1002/ente.201500304. _eprint: https://onlinelibrary.wiley.com/doi/pdf/10.1002/ente.201500304.

Thea Hauge Broholt, Michael Dahl Knudsen, and Steffen Petersen. The robustness of black and grey-box models of thermal building behaviour against weather changes. *Energy and Buildings*, 275:112460, November 2022. doi: 10.1016/j.enbuild.2022.112460.

Ion Hazyuk, Christian Ghiaus, and David Penhouet. Optimal temperature control of intermittently heated buildings using Model Predictive Control: Part I – Building modeling. *Building and Environment*, 51:379–387, May 2012a. doi: 10.1016/j.buildenv.2011.11.009.

Ion Hazyuk, Christian Ghiaus, and David Penhouet. Optimal temperature control of intermittently heated buildings using Model Predictive Control: Part II – Control algorithm. *Building and Environment*, 51:388–394, May 2012b. doi: 10.1016/j.buildenv.2011.11.008.

Tianzhen Hong, Da Yan, Simona D'Oca, and Chien-fei Chen. Ten questions concerning occupant behavior in buildings: The big picture. *Building and Environment*, 114:518–530, March 2017. doi: 10.1016/j.buildenv.2016.12.006.

Guoqing Hu and Fengqi You. Multi-zone building control with thermal comfort constraints under disjunctive uncertainty using data-driven robust model predictive control. *Advances in Applied Energy*, 9:100124, February 2023. doi: 10.1016/j.adapen.2023.100124.

Jianjun Hu and Panagiota Karava. A state-space modeling approach and multi-level optimization algorithm for predictive control of multi-zone buildings with mixed-mode cooling. *Building and Environment*, 80:259–273, October 2014. doi: 10.1016/j.buildenv.2014.05.003.

Hao Huang, Lei Chen, and Eric Hu. A neural network-based multi-zone modelling approach for predictive control system design in commercial buildings. *Energy and Buildings*, 97:86–97, June 2015. doi: 10.1016/j.enbuild.2015.03.045.

iDM Energiesysteme GmbH. AERO SLM Luftwärmepumpe - iDM Energiesysteme GmbH, 2020. URL `https://www.idm-energie.at/aero-slm-luftwaermepumpe/`.

Frank P. Incropera, editor. *Fundamentals of heat and mass transfer*. John Wiley, Hoboken, NJ, 6th ed edition, 2007. ISBN 978-0-471-45728-2.

Internation Energy Agency. Buildings, 2023a. URL `https://www.iea.org/energy-system/buildings`. (accessed: 17.02.2024).

International Energy Agency. Heating, 2023b. URL `https://www.iea.org/energy-system/buildings/heating`. (accessed: 17.02.2024).

ISO. 7730 - Ergonomics of the thermal environment — Analytical determination and interpretation of thermal comfort using calculation of the PMV and PPD indices and local thermal comfort criteria, 2005.

Kathryn B. Janda. Buildings don't use energy: people do. *Architectural Science Review*, 54(1): 15–22, February 2011. doi: 10.3763/asre.2009.0050.

Jaewan Joe, Borui Cui, Piljae Im, Jin Dong, and Kuruganti Teja. Decentralized Approach to Multi-Zone Grey-Box Modeling for Model-Based Predictive Control. In *2020 Building Performance Analysis Conference and SimBuild*, volume 9 of *SimBuild Conference*, pages 545–551. ASHRAE/IBPSA-USA, 2020.

R. E. Kalman. A New Approach to Linear Filtering and Prediction Problems. *Journal of Basic Engineering*, 82(1):35–45, March 1960. doi: 10.1115/1.3662552.

Kathryn Kaspar, Mohamed Ouf, and Ursula Eicker. A critical review of control schemes for demand-side energy management of building clusters. *Energy and Buildings*, 257:111731, February 2022. doi: 10.1016/j.enbuild.2021.111731.

Christos D. Korkas, Simone Baldi, Iakovos Michailidis, and Elias B. Kosmatopoulos. Occupancy-based demand response and thermal comfort optimization in microgrids with renewable energy sources and energy storage. *Applied Energy*, 163:93–104, February 2016. doi: 10.1016/j.apenergy.2015.10.140.

W.E. Larimore. Canonical variate analysis in identification, filtering, and adaptive control. In *29th IEEE Conference on Decision and Control*, pages 596–604 vol.2, December 1990. doi: 10.1109/CDC.1990.203665.

Yashen Lin, Timothy Middelkoop, and Prabir Barooah. Issues in identification of control-oriented thermal models of zones in multi-zone buildings. In *2012 IEEE 51st IEEE Conference on Decision and Control (CDC)*, pages 6932–6937, Maui, HI, USA, December 2012. IEEE. doi: 10.1109/CDC.2012.6425958.

Sergio Lucia, Alexandru Tătulea-Codrean, Christian Schoppmeyer, and Sebastian Engell. Rapid development of modular and sustainable nonlinear model predictive control solutions. *Control Engineering Practice*, 60:51–62, March 2017. doi: 10.1016/j.conengprac.2016.12.009.

D. Luenberger. Observers for multivariable systems. *IEEE Transactions on Automatic Control*, 11(2):190–197, April 1966. doi: 10.1109/TAC.1966.1098323. Conference Name: IEEE Transactions on Automatic Control.

Jan Lunze. *Regelungstechnik 2 : Mehrgrößensysteme, digitale Regelung, 10., überarbeitete und aktualisierte Auflage*. Springer Vieweg, Berlin, 2020. doi: https://doi.org/10.1007/978-3-662-60760-2.

Emilio T. Maddalena, Silvio A. Müller, Rafael M. Dos Santos, Christophe Salzmann, and Colin N. Jones. Experimental data-driven model predictive control of a hospital HVAC system during regular use. *Energy and Buildings*, 271:112316, September 2022. doi: 10.1016/j.enbuild.2022.112316.

H. Madsen and J. Holst. Estimation of continuous-time models for the heat dynamics of a building. *Energy and Buildings*, 22(1):67–79, March 1995. doi: 10.1016/0378-7788(94)00904-X.

Ardeshir Mahdavi, Abdolazim Mohammadi, Elham Kabir, and Lyudmila Lambeva. Occupants' operation of lighting and shading systems in office buildings. *Journal of Building Performance Simulation*, 1(1):57–65, March 2008. doi: 10.1080/19401490801906502.

O.T. Masoso and L.J. Grobler. The dark side of occupants' behaviour on building energy use. *Energy and Buildings*, 42(2):173–177, February 2010. doi: 10.1016/j.enbuild.2009.08.009.

Donald McIntyre. A guide to thermal comfort. *Applied Ergonomics*, 4(2):66–72, June 1973. doi: 10.1016/0003-6870(73)90079-3.

Daniel Minoli, Kazem Sohraby, and Benedict Occhiogrosso. IoT Considerations, Requirements, and Architectures for Smart Buildings—Energy Optimization and Next-Generation Building Management Systems. *IEEE Internet of Things Journal*, 4(1):269–283, February 2017. doi: 10.1109/JIOT.2017.2647881.

Jarinah Mohd Ali, N. Ha Hoang, M. A. Hussain, and Denis Dochain. Review and classification of recent observers applied in chemical process systems. *Computers & Chemical Engineering*, 76:27–41, May 2015. doi: 10.1016/j.compchemeng.2015.01.019.

Maximilian Mork, André Xhonneux, and Dirk Müller. Nonlinear Distributed Model Predictive Control for multi-zone building energy systems. *Energy and Buildings*, 264:112066, June 2022. doi: 10.1016/j.enbuild.2022.112066.

Sophie Naylor, Mark Gillott, and Tom Lau. A review of occupant-centric building control strategies to reduce building energy use. *Renewable and Sustainable Energy Reviews*, 96: 1–10, November 2018. doi: 10.1016/j.rser.2018.07.019.

Frauke Oldewurtel, Andreas Ulbig, Alessandra Parisio, Göran Andersson, and Manfred Morari. Reducing peak electricity demand in building climate control using real-time pricing and model predictive control. In *49th IEEE Conference on Decision and Control (CDC)*, pages 1927–1932, December 2010. doi: 10.1109/CDC.2010.5717458. ISSN: 0191-2216.

Frauke Oldewurtel, Alessandra Parisio, Colin N. Jones, Dimitrios Gyalistras, Markus Gwerder, Vanessa Stauch, Beat Lehmann, and Manfred Morari. Use of model predictive control and weather forecasts for energy efficient building climate control. *Energy and Buildings*, 45: 15–27, February 2012. doi: 10.1016/j.enbuild.2011.09.022.

Frauke Oldewurtel, David Sturzenegger, and Manfred Morari. Importance of occupancy information for building climate control. *Applied Energy*, 101:521–532, January 2013. doi: 10.1016/j.apenergy.2012.06.014.

Zhihong Pang, Yan Chen, Jian Zhang, Zheng O'Neill, Hwakong Cheng, and Bing Dong. Nationwide HVAC energy-saving potential quantification for office buildings with occupant-centric controls in various climates. *Applied Energy*, 279:115727, December 2020. doi: 10.1016/j.apenergy.2020.115727.

Gabriele Pannocchia and Mirco Calosi. Closed-Loop PARSIMonious Subspace Identification: Theory and Application to MPC. *IFAC Proceedings Volumes*, 43(5):361–366, January 2010a. doi: 10.3182/20100705-3-BE-2011.00060.

Gabriele Pannocchia and Mirco Calosi. A predictor form PARSIMonious algorithm for closed-loop subspace identification. *Journal of Process Control*, 20(4):517–524, April 2010b. doi: 10.1016/j.jprocont.2010.01.004.

June Young Park and Zoltan Nagy. Comprehensive analysis of the relationship between thermal comfort and building control research - A data-driven literature review. *Renewable and Sustainable Energy Reviews*, 82:2664–2679, February 2018. doi: 10.1016/j.rser.2017.09.102.

Theis Heidmann Pedersen and Steffen Petersen. Investigating the performance of scenario-based model predictive control of space heating in residential buildings. *Journal of Building Performance Simulation*, 11(4):485–498, July 2018. doi: 10.1080/19401493.2017.1397196.

Yuzhen Peng, Adam Rysanek, Zoltán Nagy, and Arno Schlüter. Using machine learning techniques for occupancy-prediction-based cooling control in office buildings. *Applied Energy*, 211:1343–1358, February 2018. doi: 10.1016/j.apenergy.2017.12.002.

Samuel Prívara, Jiří Cigler, Zdeněk Váňa, Frauke Oldewurtel, Carina Sagerschnig, and Eva Žáčeková. Building modeling as a crucial part for building predictive control. *Energy and Buildings*, 56:8–22, January 2013. doi: 10.1016/j.enbuild.2012.10.024.

S. Joe Qin. An overview of subspace identification. *Computers & Chemical Engineering*, 30 (10-12):1502–1513, September 2006. doi: 10.1016/j.compchemeng.2006.05.045.

Minjae Shin and Jeff S. Haberl. Thermal zoning for building HVAC design and energy simulation: A literature review. *Energy and Buildings*, 203:109429, November 2019. doi: 10.1016/j.enbuild.2019.109429.

Jan Široký, Frauke Oldewurtel, Jiří Cigler, and Samuel Prívara. Experimental analysis of model predictive control for an energy efficient building heating system. *Applied Energy*, 88(9): 3079–3087, September 2011. doi: 10.1016/j.apenergy.2011.03.009.

Andrew Sonta. Magic Walls. *Joule*, 5(8):1930–1933, August 2021. doi: 10.1016/j.joule.2021.07.018.

R. G. Steadman. The Assessment of Sultriness. Part I: A Temperature-Humidity Index Based on Human Physiology and Clothing Science. *Journal of Applied Meteorology (1962-1982)*, 18 (7):861–873, 1979. ISSN 0021-8952.

Federico Tartarini, Stefano Schiavon, Toby Cheung, and Tyler Hoyt. CBE Thermal Comfort Tool: Online tool for thermal comfort calculations and visualizations. *SoftwareX*, 12:100563, July 2020. doi: 10.1016/j.softx.2020.100563.

Charalampos Vallianos, Andreas Athienitis, and Benoit Delcroix. Automatic generation of multi-zone RC models using smart thermostat data from homes. *Energy and Buildings*, 277:112571, December 2022. doi: 10.1016/j.enbuild.2022.112571.

Peter Van Overschee and Bart De Moor. N4SID: Subspace algorithms for the identification of combined deterministic-stochastic systems. *Automatica*, 30(1):75–93, January 1994. doi: 10.1016/0005-1098(94)90230-5.

Michel Verhaegen. Subspace Techniques in System Identification. In John Baillieul and Tariq Samad, editors, *Encyclopedia of Systems and Control*, pages 1386–1396. Springer, London, 2015. doi: 10.1007/978-1-4471-5058-9_107.

Filomeno M. Vieira, Pedro S. Moura, and Aníbal T. de Almeida. Energy storage system for self-consumption of photovoltaic energy in residential zero energy buildings. *Renewable Energy*, 103:308–320, April 2017. doi: 10.1016/j.renene.2016.11.048.

G. Welch and G. Bishop. An Introduction to the Kalman Filter. 1994.

John T Wen and Sandipan Mishra. Intelligent building control systems. In *A Survey of Modern Building Control and Sensing Strategies (Advances in Industrial Control)*. Springer, 2018.

Maxwell Woody, Maryam Arbabzadeh, Geoffrey M. Lewis, Gregory A. Keoleian, and Anna Stefanopoulou. Strategies to limit degradation and maximize Li-ion battery service lifetime - Critical review and guidance for stakeholders. *Journal of Energy Storage*, 28:101231, April 2020. doi: 10.1016/j.est.2020.101231.

Jiaqing Xie, Haoyang Li, Chuting Li, Jingsi Zhang, and Maohui Luo. Review on occupant-centric thermal comfort sensing, predicting, and controlling. *Energy and Buildings*, 226:110392, November 2020. doi: 10.1016/j.enbuild.2020.110392.

Xingji Yu, Laurent Georges, Michael Dahl Knudsen, Igor Sartori, and Lars Imsland. Investigation of the Model Structure for Low-Order Grey-Box Modeling of Residential Buildings. In *Building Simulation*, pages 5076–5083, Rome, Italy, 2019. doi: 10.26868/25222708.2019.211209.